Korean Expressions Calendar

06

19

Incorrect!

<Run BTS!> Ep.150

June 19th

Incorrect!

ttaeng

When someone fails to correctly answer a quiz question, the host says, "땡 ttaeng!" It refers to the sound of a glockenspiel that is used when a quiz answer is either correct or incorrect. When the answer is correct, you can say "딩동댕 ding-dong-daeng," which is the sound of the music scale "do, mi, sol." You may recognize "땡 ttaeng" from the BTS song <DDAENG>. It uses "땡 ttaeng" in several different ways. Let's figure out which "땡 ttaeng" is the "땡 ttaeng" that means "incorrect."

07

11

ARMY! I love you.

<Run BTS!> Ep.150

July 11th

I love you.

사랑해요.

sa-rang-hae-yo

Jimin uses the expression "사랑해요 sa-rang-hae-yo" to show his love toward ARMY. "사랑해용 sa-rang-hae-yong" appears in the subtitles to describe the phrase in a cute way. If you add the batchim "ㅇ" to the last syllable in the sentence, you can make it sound even cuter. Show your love for BTS with this expression: 사랑해요 sa-rang-hae-yo!

06

20

가볍게 성공!

Succeeded easily!

<Run BTS!> Ep.104

Succeeded easily!

가볍게 성공!

ga-byeop-ge seong-gong

After Jung Kook easily succeeds in dart throwing, "**가볍게 성공** ga-byeop-ge seong-gong!" appears in the subtitles. This expression is used when something is achieved without much effort. "**가볍게** ga-byeop-ge" normally means "lightly," but here it means "simply and easily." After the phrase "**가볍게** ga-byeop-ge," you can add the word "**통과** tong-gwa," which means "pass," and say, "**가볍게 통과** ga-byeop-ge tong-gwa!" This implies that you have passed easily. How was today's Korean lesson? **가볍게 성공** ga-byeop-ge seong-gong!

07

10

<Run BTS!> Ep.110

July 10th

As expected!

역시!

yeok-shi

j-hope must look at V's drawing and guess what it is! After he correctly guesses that it is a monkey, he says "**역시** yeok-shi!" in a confident manner. This expression is used when something turns out as anticipated. BTS once posted a message on social media that read "**역시** yeok-shi! **아미가 최고야** a-mi-ga choe-go-ya," which means that ARMY is the best as expected. Why don't you also say it to them? **역시** yeok-shi! **방탄소년단이 최고야** bang-tan-so-nyeon-da-ni choe-go-ya!

06

21

If you don't play, you lose. Rock paper scissors!

<Run BTS!> Ep.127

June 21st

rock paper scissors
가위바위보
ga-wi-ba-wi-bo

RM and V are playing rock paper scissors, the simplest way to determine a winner and a loser. In Korean, rock paper scissors is called "**가위** ga-wi (scissors), **바위** ba-wi (rock), **보** bo (paper)." Before playing rock paper scissors, RM says, "**안 내면 진다** an nae-myeon jin-da," which means "You lose if you don't play." Let's try this game! **안 내면 진다** an nae-myeon jin-da, **가위바위보** ga-wi-ba-wi-bo!

07

July 9th

09

Happy Birthday, ARMY!

생일 축하해요, ARMY!

saeng-il chu-ka-hae-yo, ARMY

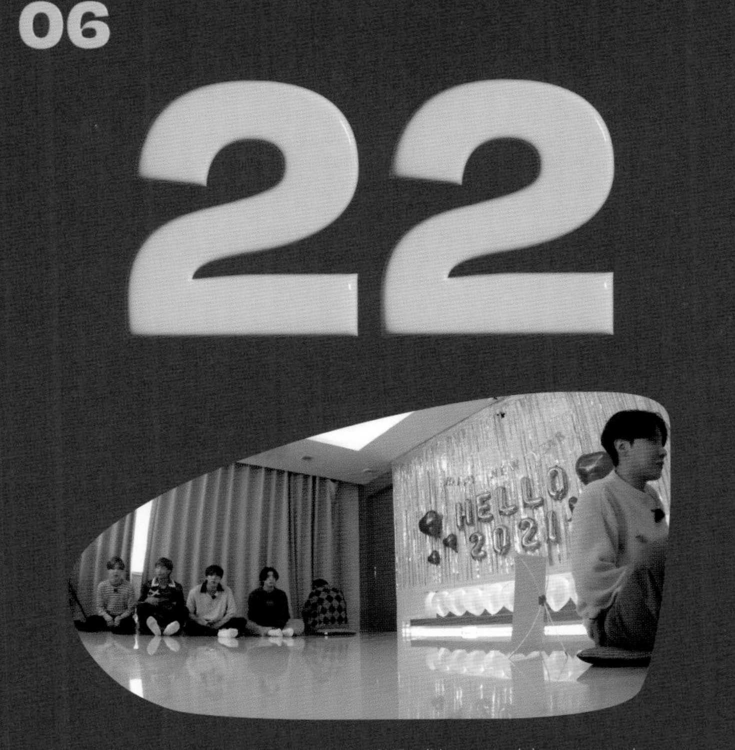

* SUGA did not attend due to personal reasons.
<Run BTS!> Ep.128

June 22nd

"it"
술래
sul-lae

There are several games in which one person is "**술래** sul-lae" ("it"). One of them is "**무궁화 꽃이 피었습니다** mu-gung-hwa kko-chi pi-eot-seum-ni-da," also known as "Red Light, Green Light." Another is "**술래잡기** sul-lae-jap-gi," where the person who is "it" must catch the remaining players. When BTS plays these kinds of games, they all run around trying to catch each other or not get caught. It's really cute! Why don't you try playing games with "**술래** sul-lae" like BTS?

07
08

우리 애
천재 아니야?

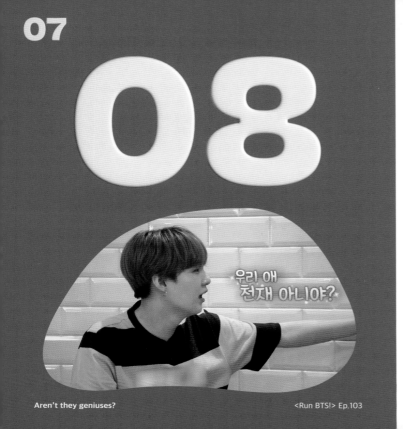

Aren't they geniuses?

<Run BTS!> Ep.103

a genius
천재
cheon-jae

RM and Jimin put some steaming *kimchi jjigae* (kimchi stew) into a squirrel-shaped bowl! When water droplets form on the squirrel, it looks like the animal is sweating. After noticing this, SUGA tells RM and Jimin that if this level of detail was intentional, they are "천재 cheon-jae," which means "genius." This word can be added after you say what someone is good at. For example, you can describe BTS as "노래 천재 no-rae cheon-jae" (singing prodigy) using the word "노래 no-rae" (song).

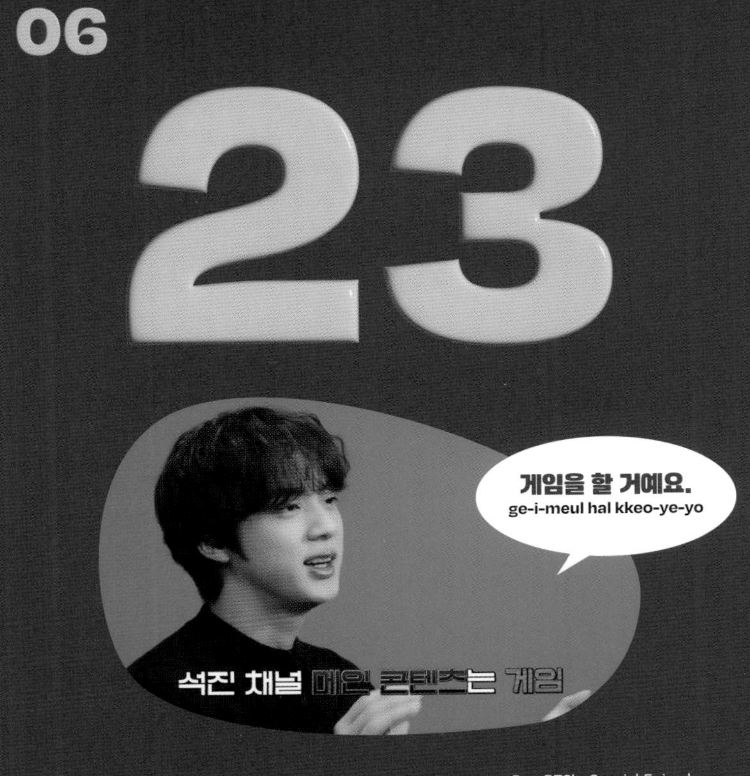

06
23
게임을 할 거예요.
ge-i-meul hal kkeo-ye-yo
석진 채널 메인 콘텐츠는 게임

The main content of Jin's channel is games.

<Run BTS!> Special Episode
- 'RUN BTS TV' On-air Part 1

I'll play games.

게임을 할 거예요.

ge-i-meul hal kkeo-ye-yo

Jin has become a temporary gaming content creator. Starting a live video stream, he says, "**게임을 할 거예요** ge-i-meul hal kkeo-ye-yo," which means "I'll play games." The phrase "**N (noun)을/를 할 거예요** eul/reul hal kkeo-ye-yo" is used to express a plan or something you will do in the future. If you want to mention your plan to study Korean, you can say "**한국어 공부를 할 거야** han-gu-geo gong-bu-reul hal kkeo-ya" or "**한국어 공부를 할 거예요** han-gu-geo gong-bu-reul hal kkeo-ye-yo," using the phrase "**한국어 공부** han-gu-geo gong-bu" (studying Korean).

* Nouns ending with a batchim use "을 eul," and nouns ending without one use "를 reul."

<Run BTS!> Ep.100

H: 재능 있어요.
jae-neung i-sseo-yo

You've got talent.
재능 있어.
jae-neung i-sseo

BTS is playing badminton using various objects as rackets! After j-hope plays well with a pot lid, SUGA compliments him by saying, "재능 있어 jae-neung i-sseo." As the word "재능 jae-neung" means "talent" or "gift," "재능 있어 jae-neung i-sseo" can be used to describe someone with talent. In polite and formal language, you can say "재능 있어요 jae-neung i-sseo-yo" to show respect. Have you already become this good at speaking Korean? 재능 있어요 jae-neung i-sseo-yo!

06

24

first place

1등

il-tteung

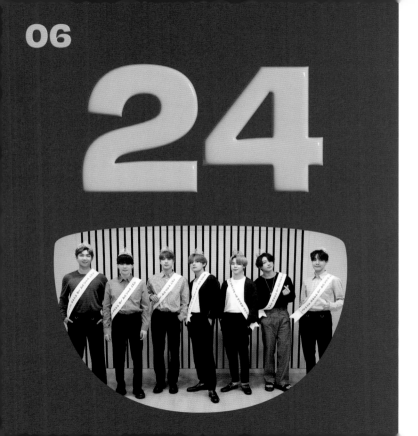

You may have heard BTS use the expression "**1등** il-tteung" when they were deciding rankings or the order in a game. "**등** deung" refers to a rank or grade and is commonly used after a number. For example, you might hear "**2등** i-deung" or "**3등** sam-deung." Use this expression when you reach the finish line first in a race: **1등** il-tteung!

07

06

Speaking of your bouncy charm

BTS (방탄소년단) j-hope's BE-hind 'Full' Story

a bouncy charm

통통 튀는 매력

tong-tong twi-neun mae-ryeok

V says that j-hope has "**통통 튀는 매력** tong-tong twi-neun mae-ryeok" when describing his personality. "**통통** tong-tong" is the sound of a small drum or a bouncy ball, and "**매력** mae-ryeok" means "charm." So "**통통 튀는 매력** tong-tong twi-neun mae-ryeok" is used to describe someone with a lively and bouncy charm. Do you know anybody who has "**통통 튀는 매력** tong-tong twi-neun mae-ryeok," like j-hope?

06

25

진짜 쉽다.
jin-jja shwip-da

초초간단

Super, super simple

<Run BTS!> Ep.125

H: 쉬워요.
shwi-wo-yo

It's easy.

쉽다.

shwip-da

After watching a famous Korean culinary researcher cook, V says, "**진짜 쉽다** jin-jja shwip-da" after realizing how simple and easy the recipe is. "**쉽다** shwip-da" is an expression used when a task is easy and simple, and "**진짜** jin-jja" is added to emphasize it. Today's Korean expression is very easy, right? Try saying "**쉽다** shwip-da" or "**쉬워요** shwi-wo-yo," if it is a piece of cake!

07

05

Because Jung Kook is cute.

<Run BTS!> Ep.129

July 5th

H: 귀엽잖아요.
gwi-yeop-ja-na-yo

Because he's cute.

귀엽잖아.

gwi-yeop-ja-na

During a tennis match, Jimin asks SUGA why Jung Kook's mistake is not marked on the scoreboard. SUGA replies, "**정국이는 귀엽잖아** jeong-gu-gi-neun gwi-yeop-ja-na," which means "Because Jung Kook is cute." "**귀엽다** gwi-yeop-da" is an expression used to describe something cute and lovely, while "**-잖아** ja-na" is used when confirming an obvious statement or situation. It means that SUGA is letting Jung Kook off the hook because he is cute. If someone asks you why BTS is your favorite band, try using the expression: **귀엽잖아** gwi-yeop-ja-na! **귀엽잖아요** gwi-yeop-ja-na-yo!

But it's difficult.

<Run BTS!> Ep.91

June 26th

H: 어려워요.
eo-ryeo-wo-yo

It's difficult.
어렵다.
eo-ryeop-da

V, who is currently hosting a quiz show, finds it more difficult to moderate the program than he expected and says, "**근데 이거 어렵다** geun-de i-geo eo-ryeop-da." "**어렵다** eo-ryeop-da" means "to be difficult." The subtitles say "**당** dang" by adding the batchim "ㅇ" to "**다** da" of "**어렵다** eo-ryeop-da." The addition of the batchim "ㅇ" makes the expression sound cute. When you have a challenging problem or question to solve, try saying "**어렵다** eo-ryeop-da" or "**어려워요** eo-ryeo-wo-yo."

Whoa, so awesome.

<Run BTS!> Ep.148

July 4th

H: 멋있어요.
meo-shi-sseo-yo

Awesome.

멋있어.

meo-shi-sseo

While taking a quiz about interior design, RM quickly gets the right answer, and j-hope says "**멋있어** meo-shi-sseo" in awe. You can use this expression to describe something or someone that is cool and awesome. In honorifics, "**멋있어요** meo-shi-sseo-yo" can be used to make it sound more formal and polite. To everyone who is studying Korean, **멋있어요** meo-shi-sseo-yo!

06

27

Now the match is tied.

<Run BTS!> Ep.129

tied score
동점
dong-jjeom

BTS splits into teams to play tennis! Jimin's team was losing 2 to 3, but his excellent serve ties the score at 3 to 3. At this moment, "**동점** dong-jjeom" (tied score) appears on the screen. "**동** dong" means "the same," and "**점** jeom" means "score." Use this expression when you score to tie a game: **동점이야** dong-jjeo-mi-ya! **동점이에요** dong-jjeo-mi-e-yo!

07

03

Handsome!

<Run BTS!> Ep.124

July 3rd

H: 잘생겼어요.
jal-saeng-gyeo-sseo-yo

Handsome.

잘생겼다.

jal-saeng-gyeot-da

Jung Kook steps onto the stage to present ideas for the content of <Run BTS!>. j-hope cheers him on by saying "**잘생겼다** jal-saeng-gyeot-da." This expression is used to describe a handsome, good-looking, or pretty person. When describing BTS or greeting them with hearty cheers, you can say "**잘생겼다** jal-saeng-gyeot-da!" or "**잘생겼어요** jal-saeng-gyeo-sseo-yo!"

06

28

Neck and neck

<Run BTS!> Ep.126

neck and neck

막상막하

mak-sang-ma-ka

V and Jung Kook are playing a game. They must use a straw to place candies on the tip of other straws. Even though it is a challenging task, they manage to do it neck-and-neck, at a similar pace. When competitors are level with one another and have an equal chance to win, it is called “**막상막하** mak-sang-ma-ka.” Try saying “**막상막하야** mak-sang-ma-ka-ya” or “**막상막하예요** mak-sang-ma-ka-ye-yo” if BTS is neck-and-neck in a game.

07

02

<Run BTS!> Ep.146

July 2nd

H: 예뻐요.
ye-ppeo-yo

Pretty.

예뻐.

ye-ppeo

In an episode of <Run BTS!>, V encounters a cat and says "예뻐 ye-ppeo" while patting the cat. This expression can be used to describe anything, including animals, people, scenery, and even ideas. When admiring BTS' gorgeous pictures or listening to their beautiful lyrics, try using this expression: 예뻐 ye-ppeo! 예뻐요 ye-ppeo-yo!

06

29

Mental concentration, once again

<Run BTS!> Special Episode
- Mini Field Day Part 1

mental concentration
정신 집중
jeong-shin jip-jung

After losing a point in a volleyball match, Jung Kook gathers himself together to concentrate. The subtitles describe this as "**정신 집중** jeong-shin jip-jung." This is a combination of "**정신** jeong-shin" (mind) and "**집중** jip-jung" (concentration). In this sense, the phrase can be used when focusing on something with your heart and soul. When you are watching a new BTS music video for the first time, you can use the expression "**정신 집중** jeong-shin jip-jung!"

07

<Weverse Live> 2021.07.19

The monsoon season seems to be coming.

장마가 올 것 같아요.

jang-ma-ga ol kkeot ga-ta-yo

In a live video stream, SUGA is talking about the weather getting hotter and adds, "**장마가 올 것 같아요** jang-ma-ga ol kkeot ga-ta-yo." In Korea, the monsoon season is called "**장마** jang-ma." As SUGA feels that "**장마** jang-ma" is coming, he says, "**장마가 올 것 같아요** jang-ma-ga ol kkeot ga-ta-yo." "**장마** jang-ma" usually occurs from the end of June to the end of July and lasts for ten days or more. If you are planning a trip that focuses on outdoor activities, it would be better not to visit Korea during this time.

06

30

<Run BTS!> will continue.

<Run BTS!> Special Episode
- Telepathy Part 2

To be continued.

계속됩니다.

gye-sok-doem-ni-da

After BTS finishes shooting an episode of <Run BTS!>, V says, "**계속됩니다** gye-sok-doem-ni-da," implying that <Run BTS!> will continue. This expression means that something will continue without end. It is commonly used at the end of an episode in a drama or webtoon. It is the equivalent of the English phrase "To be continued." Your journey of studying Korean also **계속됩니다** gye-sok-doem-ni-da!

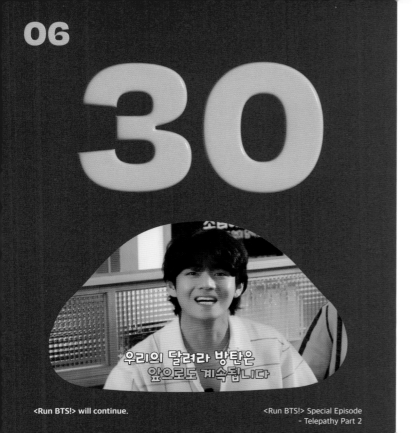

7월
chi-rwol
July

<Run BTS!> Ep.142

H: 똑똑해요!
ttok-tto-kae-yo

How smart!

똑똑해!

ttok-tto-kae

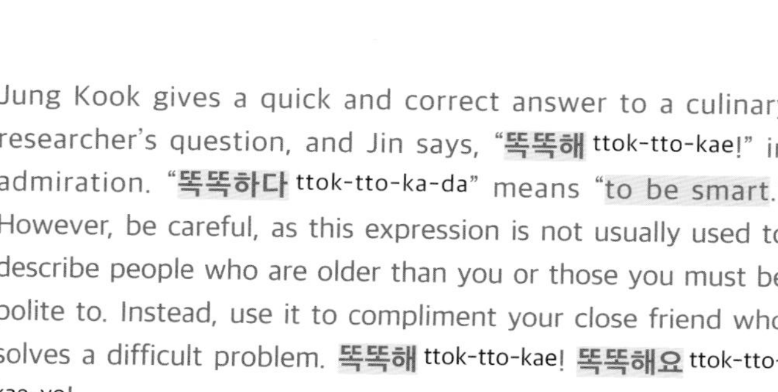

Jung Kook gives a quick and correct answer to a culinary researcher's question, and Jin says, "**똑똑해** ttok-tto-kae!" in admiration. "**똑똑하다** ttok-tto-ka-da" means "to be smart." However, be careful, as this expression is not usually used to describe people who are older than you or those you must be polite to. Instead, use it to compliment your close friend who solves a difficult problem. **똑똑해** ttok-tto-kae! **똑똑해요** ttok-tto-kae-yo!

06

18

Correct!

<Run BTS!> Ep.155

Correct!

정답!

jeong-dap

When you reply with the right answer to a quiz question, the host says, "**정답** jeong-dap!" You can also hear BTS quickly shouting "**정답** jeong-dap!" when they know the answer and then giving it. This is a compound word made of "**정** jeong" (to be correct) and "**답** dap" (answer). When someone correctly answers a Korean language question, what do you say? **정답** jeong-dap!

13

BTS (방탄소년단) 'BE' Comeback Countdown

It's breathtaking.

기가 막힙니다.

gi-ga ma-kim-ni-da

Lots of ARMY replied to postcards sent by BTS. Jin looks at a cool photo and shouts out, "**기가 막힙니다** gi-ga ma-kim-ni-da!" "**기가 막히다** gi-ga ma-ki-da" literally means that your spiritual energy (**기** gi) is blocked. Have you ever been speechless when encountering something extremely positive or negative? You can use "**기가 막히다** gi-ga ma-ki-da" in those situations. When you watch a BTS performance, you can say "**기가 막혀** gi-ga ma-kyeo!" or "**기가 막힙니다** gi-ga ma-kim-ni-da!"

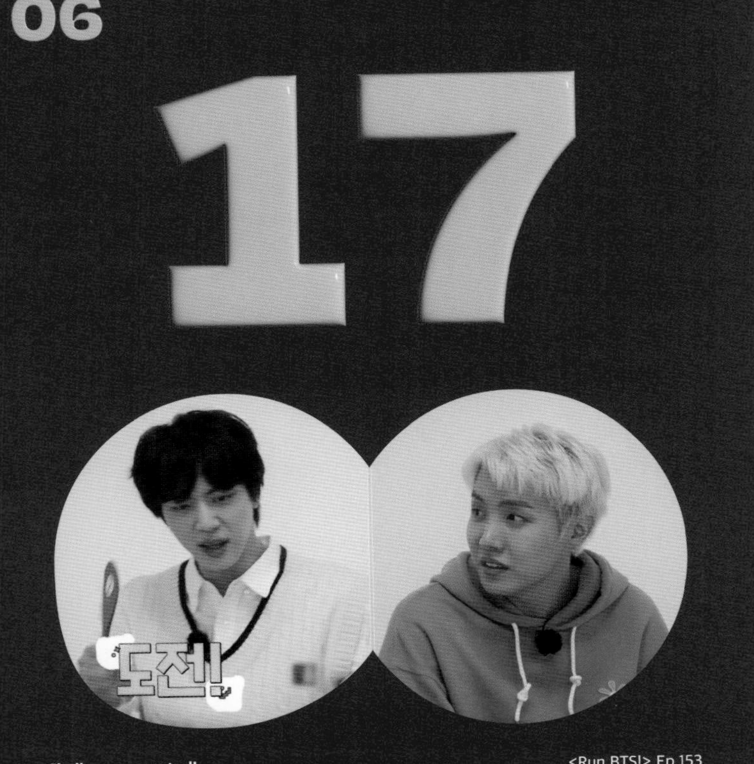

Challenge accepted!

<Run BTS!> Ep.153

Challenge accepted!

도전!

do-jeon

You can hear BTS shouting "도전 do-jeon!" before starting a game or a mission. Sometimes they start over because they did not say "도전 do-jeon!" before a challenge. This word originally means "challenge," but it is also used to express the intention to take on a challenge, just like BTS does when they shout it out loud to challenge themselves in a game or to achieve a goal. Do you want to master Korean with BTS? 도전 do-jeon!

14

That's amazing.

<Run BTS!> Ep.129

H: 신기해요.
shin-gi-hae-yo

That's amazing.
신기하다.
shin-gi-ha-da

Jimin says "**신기하다** shin-gi-ha-da" while watching V, who is swinging his tennis racket for practice. He says this because V's stance is very different from his own. When something is unusual or surprising, you can express wonder by saying this expression. When you see magic tricks that you can't figure out, you can say "**신기하다** shin-gi-ha-da" or "**신기해요** shin-gi-hae- yo."

06

16

네! 준비~ 시~ 작!

Okay! Ready, go!

<Run BTS!> Ep.123

Ready, go!

준비, 시작!

jun-bi, shi-jak

When you start an activity, you can say "**준비** jun-bi, **시작** shi-jak!" out loud. "**준비** jun-bi" means "preparation," and "**시작** shi-jak" means "start." It is commonly used with activities that have a time limit. When the time is up, you can say "**끝** kkeut" to signal the end. Are you ready to study Korean with BTS today? If so, **준비** jun-bi, **시작** shi-jak!

15

Which BTS performance, as voted by ARMY, is ranked as the best of all time?

<Run BTS!> Ep.144

the best performance
최고의 무대
choe-go-ui mu-dae

In a quiz, BTS has to guess what ARMY chose as their best performance! "The best performance" can be translated as "**최고의 무대** choe-go-ui mu-dae" in Korean. When talking about something that is the best, you can say "**최고의** choe-go-ui N (noun)." For example, the best song (**노래** no-rae) can be called "**최고의 노래** choe-go-ui no-rae," and the best album (**앨범** ael-beom) can be called "**최고의 앨범** choe-go-ui ael-beom."

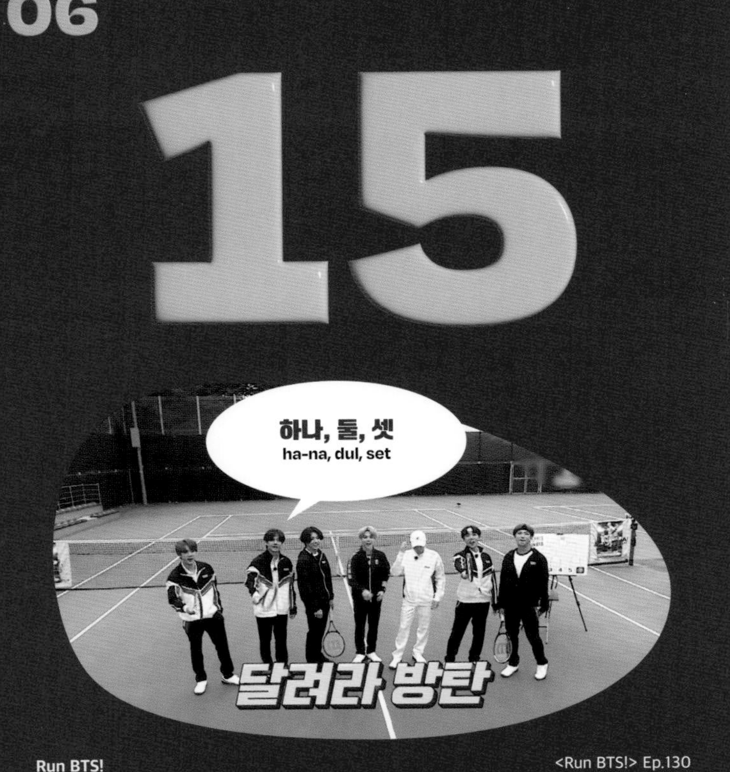

one, two, three

하나, 둘, 셋

ha-na, dul, set

To help everyone chant "Run BTS!" at the same time, V shouts "**하나** ha-na, **둘** dul, **셋** set." "**하나** ha-na," "**둘** dul," and "**셋** set" are the cardinal numbers in Korean, meaning 1, 2, and 3, respectively. People sometimes say only "**둘** dul, **셋** set" without "**하나** ha-na." Does it sound familiar? Yes, you're probably thinking of BTS' energetic greeting, which is "**둘** dul, **셋** set" followed by "**방** bang! **탄** tan!"

16

I'm rather good at this.

<Run BTS!> Ep.128

H: 저 좀 잘해요.
jeo jom jal-hae-yo

I'm rather good at this.
나 좀 잘해.
na jom jal-hae

Jimin is about to play a game where the others have to guess which song he is playing on the harmonica! He says he's good at this game and adds, "나 좀 잘해 na jom jal-hae." This expression is combined with "좀 jom" (a bit) and "잘하다 jal-ha-da" (to do well). When you think you're skilled at something, you can say "나 좀 잘해 na jom jal-hae" or "저 좀 잘해요 jeo jom jal-hae-yo."

06

14

ARMY, I purple you.

<Run BTS!> Ep.97

H: 보라해요.
bo-ra-hae-yo

I purple you.

보라해.

bo-ra-hae

Do you know the origin of "보라해 bo-ra-hae," which BTS and ARMY use instead of "사랑해 sa-rang-hae?" (👉 July 11th) "보라 bo-ra" means "purple," and "보라해 bo-ra-hae" was created by V when he saw ARMY BOMBs covered with purple plastic bags at a fan meeting in 2016. Just like purple is the color at the end of a rainbow, V explained, the love and trust that BTS and ARMY have for each other will be there in the end. 보라해 bo-ra-hae♥

17

Ignorance is bliss.

모르는 게 상책

mo-reu-neun ge sang-chaek

In a live video stream, RM is asked to choose between "모르는 게 상책 mo-reu-neun ge sang-chaek" and "모르는 개 산책 mo-reu-neun gae san-chaek." This is one of those "would you rather" questions that were once popular in Korea using two phrases with similar pronunciation. The former means "ignorance is bliss" and the latter means "walking a dog (개 gae) you don't know." While the two phrases are irrelevant, RM chooses "모르는 게 상책 mo-reu-neun ge sang-chaek." Do you agree that sometimes in life, not knowing is better? If so, try using this expression: 모르는 게 상책 mo-reu-neun ge sang-chaek.

06

13

Happy Birthday, BTS!

생일 축하해요, BTS!

saeng-il chu-ka-hae-yo, BTS

18

<Run BTS!> Ep.105

H: 딱이네요.
tta-gi-ne-yo

It hits the spot.

딱이네.

tta-gi-ne

SUGA is making modifications to his clothes in an episode of <Run BTS!>. He likes what he has done because it resembles the season's trends and says, "**딱이네** tta-gi-ne." The word "**딱** ttak" refers to something that fits perfectly or hits the spot. When things fall into place or suit someone well, you can use the expression "**딱이네** tta-gi-ne." In honorifics, you can say "**딱이네요** tta-gi-ne-yo" to make it sound formal and polite. Are you studying Korean with <365 BTS DAYS>? **딱이네요** tta-gi-ne-yo!

12

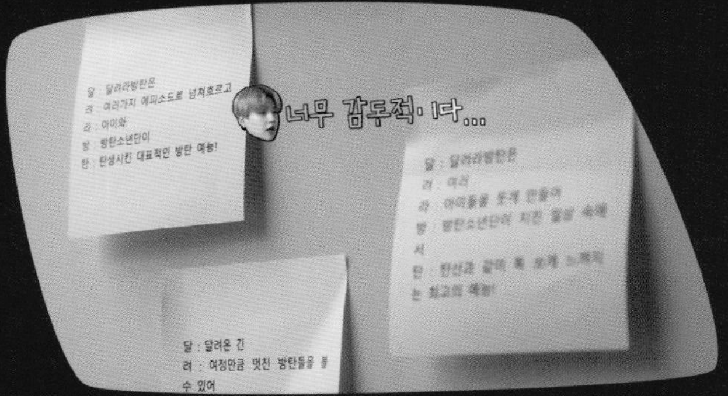

It's very touching...

<Run BTS!> Ep.154

H: 감동적이에요.
gam-dong-jeo-gi-e-yo

It's touching.

감동적이다.

gam-dong-jeo-gi-da

SUGA is moved by the five-line acrostic poems written by ARMY using the Korean spelling of <Run BTS!>, which is "**달려라 방탄** dal-lyeo-ra bang-tan," and expresses his emotions by saying "**감동적이다** gam-dong-jeo-gi-da." You can use this expression when you are moved by something such as a song or writing. Are you moved by the lyrics written by BTS for ARMY? Then try saying "**감동적이다** gam-dong-jeo-gi-da" or "**감동적이에요** gam-dong-jeo-gi-e-yo."

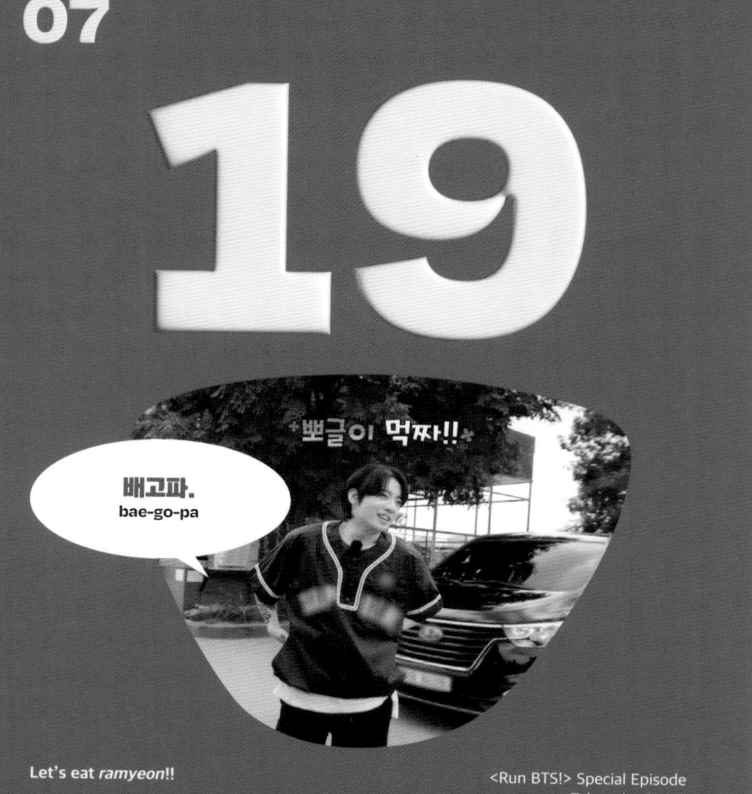

Let's eat *ramyeon*!!

<Run BTS!> Special Episode - Telepathy Part 2

H: 배고파요.
bae-go-pa-yo

I'm hungry.
배고파.
bae-go-pa

After arriving at Hangang Park, Jung Kook suggests eating *ramyeon* (Korean instant noodles), the ultimate picnic food on the riverside in Korea. He says "**배고파** bae-go-pa," which means "I'm hungry." On the other hand, if you are full, you can say "**배불러(요)** bae-bul-leo(yo)" (👉 July 27th). When you are hungry, try saying "**배고파(요)** bae-go-pa(yo)."

Burning with determination

<Run BTS!> Ep.134

burning with determination

의지 활활

ui-ji hwal-hwal

In a game of guessing a song title while watching a staff member dance, j-hope stares at the screen with great determination to win. The subtitles describe him as "의지 활활 ui-ji hwal-hwal." "활활 hwal-hwal" refers to the shape of a fiercely blazing fire. So "의지 활활 ui-ji hwal-hwal" means that someone is eager to do something, like a burning fire. When you are determined to finish your work on time so that you can leisurely watch BTS video clips in the evening, you can describe yourself as "의지 활활 ui-ji hwal-hwal."

07

20

맛있게 드세요!

Enjoy your food!

<Run BTS!> Ep.133

Enjoy your food.

맛있게 드세요.

ma-shit-ge deu-se-yo

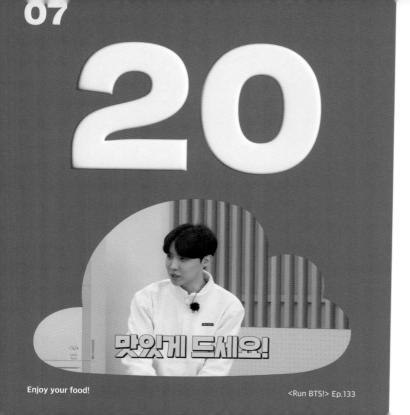

Before eating snacks together, j-hope says, "**맛있게 드십시오** ma-shit-ge deu-ship-shi-o" to the others. This expression, which means "Enjoy your food," is a polite expression used before eating food. It originated from the phrase "**맛있게 드세요** ma-shit-ge deu-se-yo," which is used more often. And you can casually say "**맛있게 먹어** ma-shit-ge meo-geo" between close friends.

10

<Weverse Live> 2021.10.20

get better

갈수록 좋아지다

gal-ssu-rok jo-a-ji-da

Jin has made a promise to exercise whenever he visits the office. As a result, his frequent visits to the office have made him work out every day. He uses the expression, "**몸이 갈수록 좋아지다** mo-mi gal-ssu-rok jo-a-ji-da," which implies that he has gotten into better shape over time. "**갈수록 좋아지다** gal-ssu-rok jo-a-ji-da" means that the condition of something improves as time goes by. If you have had a cold but have started feeling better over time, you can say "**갈수록 좋아져** gal-ssu-rok jo-a-jeo" or "**갈수록 좋아져요** gal-ssu-rok jo-a-jeo-yo."

Thank you for the meal.

<Run BTS!> Ep.154

Thank you for the meal.

잘 먹겠습니다.

jal meok-get-seum-ni-da

Jimin is sitting in front of a dining table full of delicacies. He says "**잘 먹겠습니다** jal meok-get-seum-ni-da" before starting his meal. This expression literally means "I will eat well," and is used as a greeting before meals. After finishing a dish, you can say "**잘 먹었습니다** jal meo-geot-seum-ni-da" (☛ July 28th), which means "I really enjoyed the food."

09

June 9th

<Weverse Live> 2021.01.19

I'm into art history.

미술사에 빠져 있습니다.

mi-sul-ssa-e ppa-jeo it-seum-ni-da

When discussing his latest interest in a live video stream, RM says, "**미술사에 빠져 있습니다** mi-sul-ssa-e ppa-jeo it-seum-ni-da." "**빠지다** ppa-ji-da" typically means "to fall into a hole or a puddle." When you say "**N (noun)에 빠져 있습니다** e ppa-jeo it-seum-ni-da," it indicates that you are highly interested in something. You can describe yourself as being into BTS using this expression: **방탄소년단에 빠져 있어** bang-tan-so-nyeon-da-ne ppa-jeo i-sseo, **방탄소년단에 빠져 있습니다** bang-tan-so-nyeon-da-ne ppa-jeo it-seum-ni-da.

07

[2020 FESTA] BTS (방탄소년단) '방탄생파'

July 22nd

H: 진짜 맛있어요.
jin-jja ma-shi-sseo-yo

Really delicious.

진짜 맛있어.

jin-jja ma-shi-sseo

At the BTS birthday party, Jung Kook has some *miyeokguk* (seaweed soup) and says, "**진짜 맛있어** jin-jja ma-shi-sseo." When eating something delicious, you can say "**맛있어(요)** ma-shi-sseo(yo)." In the dialogue, Jung Kook adds the phrase "**진짜** jin-jja" to emphasize how good the soup is. Instead of this, you can use similar expressions such as "**정말** jeong-mal," "**엄청** eom-cheong," or "**너무** neo-mu," or use more informal expressions such as "**완전** wan-jeon" or "**대박** dae-bak" instead.

Since it's hot, let's have ice cream—just one.

<Run BTS!> Special Episode
- Telepathy Part 2

H: 하나만 먹어요.
ha-na-man meo-geo-yo

Let's eat just one.

하나만 먹자.

ha-na-man meok-ja

Upon arriving at Hangang Park, Jung Kook suggests that they get only one ice cream since the weather is hot by saying, "**하나만 먹자** ha-na-man meok-ja." "**하나** ha-na" means "one," and "**-자** ja" is used to close a sentence when recommending or suggesting something to close friends or someone younger than you. If you go shopping with your friend and want to buy a bag of snacks, you can use "**사다** sa-da" (to buy) and say, "**하나만 사자** ha-na-man sa-ja."

23

바삭 불호파

Not a fan of crispy food

<Run BTS!> Ep.142

July 23rd

crispy

바삭(바삭)

ba-sak(ba-sak)

"**바삭** ba-sak" and "**바삭바삭** ba-sak-ba-sak" are the sounds made when eating something crispy, such as fried chicken or potato chips. You can say, "**바삭(바삭)해** ba-sak(ba-sak)hae" or "**바삭(바삭)해요** ba-sak(ba-sak)hae-yo" when you are enjoying crispy food. Jin says that he is not fond of crispy food. What about you? Do you like crispy food?

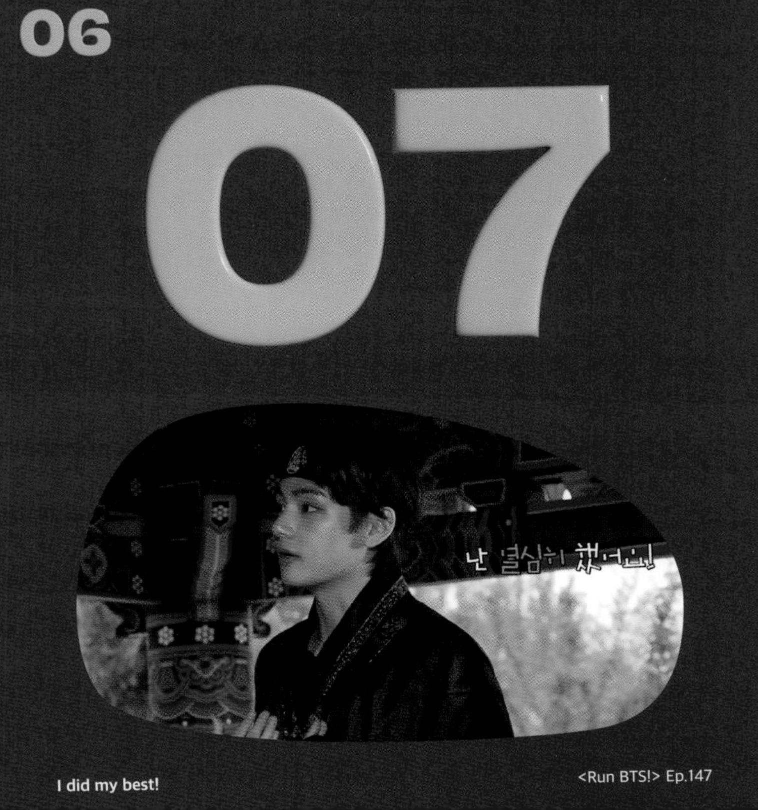

06
07
난 열심히 했어요
I did my best!
<Run BTS!> Ep.147

I did my best.

열심히 했어요.

yeol-sshim-hi hae-sseo-yo

V is suspected of being a thief in a game, but he defends himself by saying, "열심히 했어요 yeol-sshim-hi hae-sseo-yo," which implies that he did his best to accomplish the mission. This expression is used when telling someone that you tried your best. Did you study Korean hard today? If so, try using the expression "열심히 했어 yeol-sshim-hi hae-sseo" or "열심히 했어요 yeol-sshim-hi hae-sseo-yo."

24

I love you.

<Run BTS!> Ep.153

H: 달콤해요.
dal-kom-hae-yo

Sweet.

달콤하다.

dal-kom-ha-da

After listening to Jung Kook sing, j-hope admires his voice by saying "**달콤하다** dal-kom-ha-da." This expression is originally used to describe a sweet flavor, but it can also be used to describe a person or thing that makes you happy, such as a love confession or a pleasant nap. Try saying this expression when listening to BTS' warm voices: **달콤하다** dal-kom-ha-da! **달콤해요** dal-kom-hae-yo!

June 6th

06

06

슈가
좋은 취미가 하나 생긴 거 같은데요?

SUGA: It seems like you've found a good new hobby.

<Run BTS!> Ep.99

find a new hobby

취미가 생기다

chwi-mi-ga saeng-gi-da

BTS has learned how to arrange flowers, and RM, in particular, has taken a liking to it. SUGA tells him that he thinks RM has a new hobby using the expression "**취미가 생기다** chwi-mi-ga saeng-gi-da." This implies that someone has developed a new hobby, as "**취미** chwi-mi" means "hobby" and "**생기다** saeng-gi-da" means "to be formed." If you have recently acquired a new hobby, you can use the expression "**취미가 생겼어** chwi-mi-ga saeng-gyeo-sseo," or "**취미가 생겼어요** chwi-mi-ga saeng-gyeo-sseo-yo."

Childlike palate /
It looks spicy.

<Run BTS!> Special Episode
- 'RUN BTS TV' On-air Part 2

H: 매울 것 같아요.
mae-ul kkeot ga-ta-yo

It looks spicy.
매울 것 같아.
mae-ul kkeot ga-ta

When Jimin shows a bottle of spicy-looking sauce during a live video stream, V says "**매울 것 같아** mae-ul kkeot ga-ta." This expression conjugates "**맵다** maep-da" (to be spicy) and is used when you think some food might be spicy. Does the photo of the recommended dish at the restaurant look red and spicy? Then try saying "**매울 것 같아** mae-ul kkeot ga-ta" or "**매울 것 같아요** mae-ul kkeot ga-ta-yo."

H: 할 거예요? 말 거예요?
hal kkeo-ye-yo? mal kkeo-ye-yo?

Are you going to do it or not?

할 거야? 말 거야?

hal kkeo-ya? mal kkeo-ya?

During a dice game, Jung Kook asks V, "할 거야 hal kkeo-ya? 말 거야 mal kkeo-ya?" because V starts to raise an objection but then hesitates. Jung Kook is asking V whether he is going to voice his objection or not. This phrase is commonly used when asking about someone's intentions or asking someone to choose whether to do something or not. For instance, if your close friend hesitates before bungee jumping, you can ask them "할 거야 hal kkeo-ya? 말 거야 mal kkeo-ya?"

07

26

[VLOG] 제이홉 | 베리임폴턴트비지니스

a brunch

아침 겸 점심

a-chim gyeom jeom-shim

j-hope, who eats his first meal around 12:20 p.m., describes it as "**아침 겸 점심** a-chim gyeom jeom-shim" (brunch) because it is too late to call it breakfast. This expression can be shortened to the new word "**아점** a-jeom" these days. Similarly, you might hear "**점심 겸 저녁** jeom-shim gyeom jeo-nyeok," which refers to a meal that serves both lunch and dinner at the same time, and this is also shortened to "**점저** jeom-jeo."

<Weverse Live> 2020.03.26

H: 오늘 무슨 요일이에요?
o-neul mu-seun yo-i-ri-e-yo

What day is it today?

오늘 무슨 요일이지?

o-neul mu-seun yo-i-ri-ji

During a live video stream, Jimin tries to remember what day it is and asks himself, "**오늘 무슨 요일이지** o-neul mu-seun yo-i-ri-ji?" When asking directly about the day of the week, you can say "**오늘 무슨 요일이야** o-neul mu-seun yo-i-ri-ya?" or "**오늘 무슨 요일이에요** o-neul mu-seun yo-i-ri-e-yo?" to the other person. If you're alone and curious about the day of the week, try saying to yourself, "**오늘 무슨 요일이지** o-neul mu-seun yo-i-ri-ji?"

07

27

배불러

I'm full.

<Run BTS!> Special Episode
- Telepathy Part 2

H: 배불러요.
bae-bul-leo-yo

I'm full.

배불러.

bae-bul-leo

BTS is having a meat feast! At the end of the meal, Jin says, "**배불러** bae-bul-leo." After eating, when your stomach (**배** bae) is full, you can say this expression. And when you are hungry, you can say "**배고파(요)** bae-go-pa(yo)" (☛ July 19th) as you learned. If someone offers you something to eat when you're already full, try saying "**배불러(요)** bae-bul-leo(yo)."

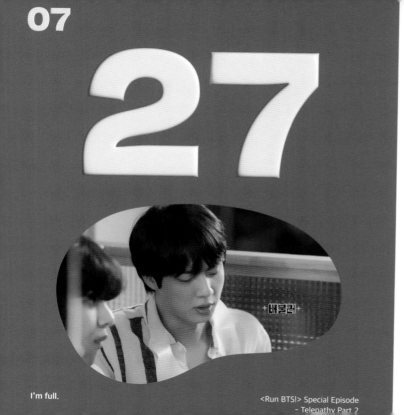

06

03

그럼 안 할래ㅋㅋ

Then I won't do it.

<Run BTS!> Special Episode – Next Top Genius Part 2

June 3rd

H: 안 할래요.
an hal-lae-yo

I won't do it.

안 할래.

an hal-lae

BTS is playing a dice game! V is about to protest during Jung Kook's turn, but he suddenly changes his mind, saying, "안 할래 an hal-lae." The phrase "-(으)ㄹ래 (eu)l-lae" is used to express willingness. You can express unwillingness by adding "안 an," which has a negative meaning. When asked if you will have breakfast and you won't, you can say "안 먹을래 an meo-geul-lae" or "안 먹을래요 an meo-geul-lae-yo," using "먹다 meok-da" (to eat).

28

I really enjoyed the food.

잘 먹었어요.

jal meo-geo-sseo-yo

BTS, who meets after a long time for the filming of <Run BTS!> is talking about their current situations, and j-hope thanks Jin for the watermelon he gave him as a gift, saying "수박 잘 먹었어요 su-bak jal meo-geo-sseo-yo." "잘 먹었어요 jal meo-geo-sseo-yo" is commonly used after a meal to thank the person who prepared or provided it. For a more polite expression, you can say "잘 먹었습니다 jal meo-geot-seum-ni-da." Between close friends, you can simply say "잘 먹었어 jal meo-geo-sseo."

06
02
자 여러분 우리 차분히 생각합시다

Everyone, let's calm down and think.

<Run BTS!> Ep.131

Let's calm down and think.

차분히 생각합시다.

cha-bun-hi saeng-ga-kap-shi-da

BTS is in the middle of a game where players are penalized when they do specific actions. After being punished many times in a row, they try to guess what the prohibited actions are. At that moment, j-hope says, "차분히 생각합시다 cha-bun-hi saeng-ga-kap-shi-da," which means "Let's calm down and think." "차분히 cha-bun-hi" means "calmly," and "생각하다 saeng-ga-ka-da" means "to think." If you get lost while traveling in Korea with someone, you can say "차분히 생각하자 cha-bun-hi saeng-ga-ka-ja," or "차분히 생각합시다 cha-bun-hi saeng-ga-kap-shi-da."

07

Time for a return of <Run BTS!> toast

<Run BTS!> Special Episode
- Telepathy Part 1

make a toast

건배(를) 하다

geon-bae(reul) ha-da

BTS has gathered for <Run BTS!> and is making a toast. This is called "건배(를) 하다 geon-bae(reul) ha-da" in Korean. "건배 geon-bae" means "to raise your glass and wish others the best while drinking." When making a toast or proposing a drink, we can deliver a message through "건배사 geon-bae-sa," which refers to this message in Korean. Let's learn about a specific kind of "건배사 geon-bae-sa" tomorrow!

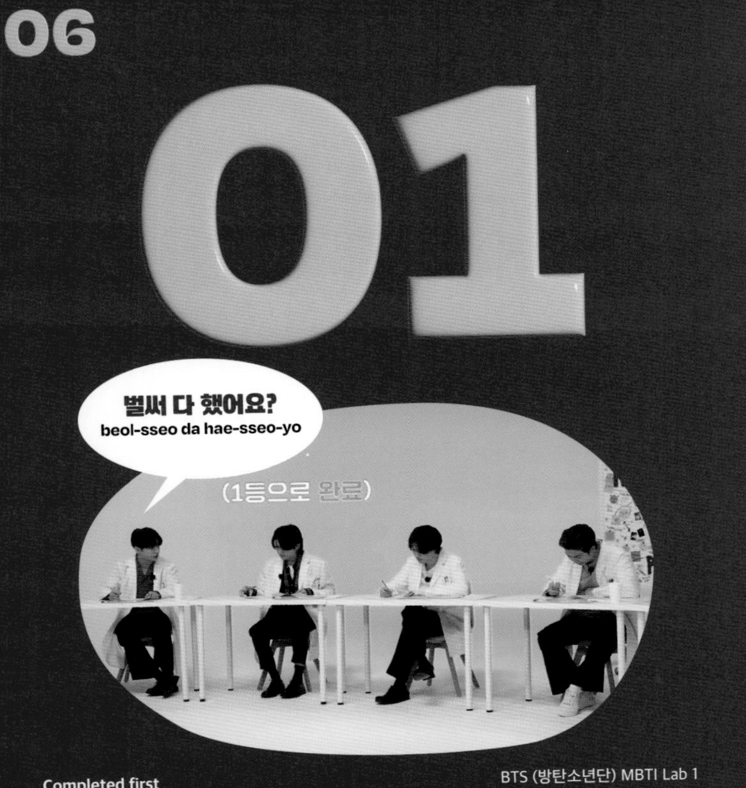

06
01
벌써 다 했어요?
beol-sseo da hae-sseo-yo
(1등으로 완료)
Completed first
BTS (방탄소년단) MBTI Lab 1

Are you done already?

벌써 다 했어요?

beol-sseo da hae-sseo-yo

When SUGA finishes a personality test first, Jung Kook says, "벌써 다 했어요 beol-sseo da hae-sseo-yo?" which means "Are you done already?" "벌써 beol-sseo" (already) is used when something is completed faster than expected. If your friend, who is learning Korean with you, finishes studying earlier than you expected, you can say "벌써 다 했어 beol-sseo da hae-sseo?" or "벌써 다 했어요 beol-sseo da hae-sseo-yo?"

07

30

<Weverse Live> 2021.11.22

Cheers!

위하여!

wi-ha-yeo

BTS is preparing a toast to celebrate winning a music award! RM proposes a toast of "아미를 위하여 a-mi-reul wi-ha-yeo," which means "Cheers to ARMY." "위하여 wi-ha-yeo," which is one of "건배사 geon-bae-sa" (☛ July 29th), is a toast used when drinking alcohol. While it can be used alone, it is also possible to add a person's name or words such as "love" and "friendship," as in "사랑을 위하여 sa-rang-eul wi-ha-yeo" (Cheers to love). Shall we have a toast to BTS this time? 방탄소년단을 위하여 bang-tan-so-nyeon-da-neul wi-ha-yeo!

6월

yu-wol

[2020 FESTA] BTS (방탄소년단) '방탄생파'

clink

짠

jjan

Have you seen BTS clink glasses when celebrating their birthdays or after winning an award? When they do, they say, "짠 jjan." This is the sound of glasses clinking during a toast. Instead of saying "**위하여** wi-ha-yeo" (☛ July 30th) for a toast, you can simply say "**짠** jjan." Let's watch a video clip of BTS having a toast. Why don't you clink glasses together? 짠 jjan!

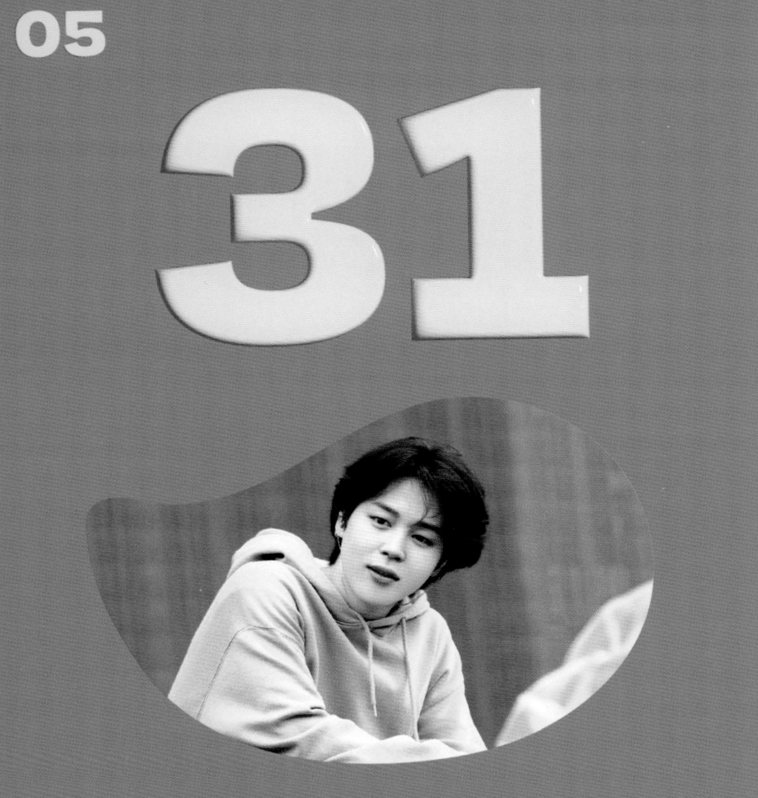

There's a lot to do.

할 거 많은데?

hal kkeo ma-neun-de

Jimin is watching Jin and SUGA cook. He is surprised by how many steps there are and says, "할 거 많은데 hal kkeo ma-neun-de?" This sentence is a combination of "할 거 많다 hal kkeo man-ta," which means "There's a lot to do," and "-은데/는데 eun-de/neun-de?" a suffix used to express admiration or wonder about something. Before starting to study or work, if there are too many things that you need to take care of, you can try saying "할 거 많은데 hal kkeo ma-neun-de?" or "할 거 많은데요 hal kkeo ma-neun-de-yo?" to express that you feel overwhelmed or surprised by the amount of work that needs to be done.

* Use "은eun" if the previous word ends with a batchim and "는neun" otherwise.

8월

pa-rwol

AUGUST

Jung Kook: I dropped my cell phone.
<Run BTS!> Ep.91

H: 휴대폰을 떨어뜨렸어요.
hyu-dae-po-neul tteo-reo-tteu-ryeo-sseo-yo

I dropped my cell phone.

휴대폰을 떨어뜨렸어.

hyu-dae-po-neul tteo-reo-tteu-ryeo-sseo

Have you ever dropped your cell phone (휴대폰 hyu-dae-pon) and cracked its screen? It's such an awful feeling. When you drop your phone, you can say, "휴대폰을 떨어뜨렸어 hyu-dae-po-neul tteo-reo-tteu-ryeo-sseo," just like Jung Kook says. This expression conjugates "떨어뜨리다 tteo-reo-tteu-ri-da" (to drop). In honorific form, you can say "휴대폰을 떨어뜨렸어요 hyu-dae-po-neul tteo-reo-tteu-ryeo-sseo-yo." Take care of your belongings, and make sure not to drop anything!

01

아우 더워

Oh, it's hot.

<Run BTS!> Special Episode
- 'RUN BTS TV' On-air Part 2

August 1st

H: 더워요.
deo-wo-yo

It's hot.

deo-wo

When Jung Kook feels hot while filming <Run BTS!>, he says "더워 deo-wo." This expression is used to describe hot weather. Meanwhile, when it is cold, you can say "추워 chu-wo," which is used to describe cold weather. The basic forms of these expressions are "덥다 deop-da" and "춥다 chup-da" respectively. If someone asks you about the weather on a hot day, you can answer like this: 더워 deo-wo, 더워요 deo-wo-yo.

05

29

Obviously, it seems to be taking some time.

<Run BTS!> Ep.126

H: 오래 걸리겠는데요?
o-rae geol-li-gen-neun-de-yo

This will take some time.
오래 걸리겠는데?
o-rae geol-li-gen-neun-de

Jin finds it hard to bounce a ping-pong ball into a cup to complete his mission in a game. At that moment, RM says, "오래 걸리겠는데 o-rae geol-li-gen-neun-de?" This expression is used when something takes time to do, and "-겠는데 gen-neun-de?" is added as a suffix to show speculation. As such, it means that something seems to take quite some time to finish. If you are heading somewhere and it seems to be taking a while because of a traffic jam, try saying "오래 걸리겠는데 o-rae geol-li-gen-neun-de?" or "오래 걸리겠는데요 o-rae geol-li-gen-neun-de-yo?"

08

02

BTS (방탄소년단) MBTI Lab 2

Hello! (Goodbye!)

an-nyeong

BTS waves to ARMY and says, "안녕 an-nyeong!" As the most basic casual greeting, "안녕 an-nyeong" can be used to say both hello and goodbye. If you need to use honorifics in a formal setting, you can say "안녕하세요 an-nyeong-ha-se-yo" when meeting and "안녕히 가세요 an-nyeong-hi ga-se-yo" when parting, bending slightly forward at the waist. Let's greet someone we meet on the street cheerfully. 안녕 an-nyeong! 안녕하세요 an-nyeong-ha-se-yo!

05

28

Time flies faster than expected.

<Run BTS!> Ep.103

H: 시간이 빨리 가네요.
shi-ga-ni ppal-li ga-ne-yo

Time flies.

시간이 빨리 가네.

shi-ga-ni ppal-li ga-ne

RM and Jimin have become avatar chefs for SUGA! Although they haven't done much yet, j-hope says that 10 minutes have already passed. Jin then says, "시간이 빨리 가네 shi-ga-ni ppal-li ga-ne," because he is surprised at how fast the time is going. When you are busy with a heavy workload, have you ever looked up and realized that time has flown by? If so, you can say "시간이 빨리 가네 shi-ga-ni ppal-li ga-ne," or "시간이 빨리 가네요 shi-ga-ni ppal-li ga-ne-yo."

08

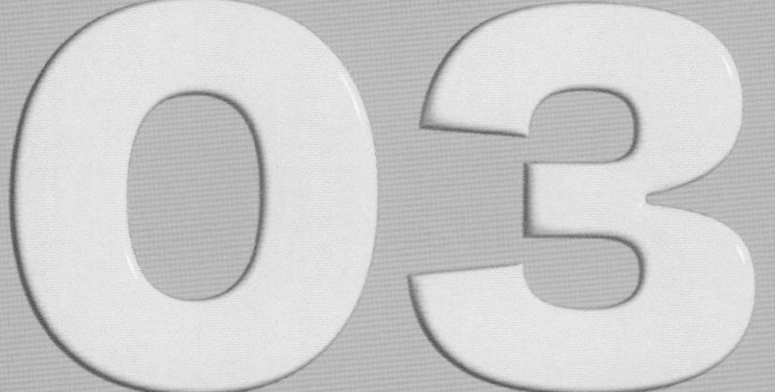

August 3rd

Welcome.

환영합니다.

hwa-nyeong-ham-ni-da

j-hope greets ARMY who has joined a live video stream by saying, "환영합니다 hwa-nyeong-ham-ni-da!" This expression is used when welcoming someone. You can often hear hosts saying this expression to greet guests at airports, hotels, concerts, and festivals. To every ARMY who has just started learning Korean, 환영합니다 hwa-nyeong-ham-ni-da!

05

27

V
저는 (명절 음식으로) 집 밥 추천합니다

I recommend homemade meals (for the holidays).

[EPISODE] BTS (방탄소년단) 2021 'DALMAJUNG' Shoot

I recommend it.

추천합니다.

chu-cheon-ham-ni-da

As V usually enjoys homemade meals during traditional holidays, he naturally replies with home-cooked meals when asked to recommend some holiday dishes. When you recommend something, you can use the expression "추천합니다 chu-cheon-ham-ni-da." If someone has just become a BTS fan and asks for a song recommendation, you can say its title and add "추천해 chu-cheon-hae," or "추천합니다 chu-cheon-ham-ni-da."

08

04

August 4th

Nice to meet you.

반가워요.

ban-ga-wo-yo

V greets Jimin during a live video stream by saying, "반가워요 ban-ga-wo-yo!" This is a casual greeting often used when meeting someone for the first time, unexpectedly, or after a long time. To every ARMY who has just started learning Korean, 반가워요 ban-ga-wo-yo!

05

26

Are you laughing? Come here.

<Run BTS!> Ep.128

H: 일로 오세요.
il-lo o-se-yo

Come here.

일로 와.

il-lo wa

Jin has become "it" in a game of "**무궁화 꽃이 피었습니다** mu-gung-hwa kko-chi pi-eot-seum-ni-da," also known as "Red Light, Green Light." In this game, any players still moving when the person who is "it" turns around must go to their side. When Jin notices Jimin's smile and slight movement, he says, "**일로 와** il-lo wa," meaning "Come here." "**일로** il-lo" is short for "**이리로** i-ri-ro," which means "here." Why don't we try a game of "**무궁화 꽃이 피었습니다** mu-gung-hwa kko-chi pi-eot-seum-ni-da," as BTS does? When you become "it" and see someone move, try saying "**일로 와** il-lo wa," or "**일로 오세요** il-lo o-se-yo."

08

How have you been?

<Run BTS!> Special Episode
- Telepathy Part 1

How have you been?

잘 지내셨어요?

jal ji-nae-shyeo-sseo-yo

BTS starts filming <Run BTS!> again after a long break. They ask each other how they have been doing by saying, "잘 지내셨어요 jal ji-nae-shyeo-sseo-yo?" This expression is used when you ask someone you haven't seen in a while if they have been doing well during the time you were apart. You can casually say "잘 지냈어 jal ji-nae-sseo?" to your close friends. Try talking to someone you haven't seen for a long time. 잘 지냈어 jal ji-nae-sseo? 잘 지내셨어요 jal ji-nae-shyeo-sseo-yo?

25

I couldn't practice tennis!

<Run BTS!> Ep.130

H: 연습 못 했어요.
yeon-seup mot hae-sseo-yo

I couldn't practice.

연습 못 했다.

yeon-seup mot haet-da

BTS gets together to play tennis! With a bright smile, Jung Kook says, "테니스 연습 못 했다 te-ni-seu yeon-seup mot haet-da," which means he couldn't practice tennis. In this expression, "연습 yeon-seup" means "practice" and "못 했다 mot haet-da" means that you couldn't do something before. If you were too busy to study Korean yesterday, you could use the phrase "한국어 공부 han-gu-geo gong-bu" (studying Korean) and make sentences like this: 한국어 공부 못 했다 han-gu-geo gong-bu mot haet-da, 한국어 공부 못 했어요 han-gu-geo gong-bu mot hae-sseo-yo.

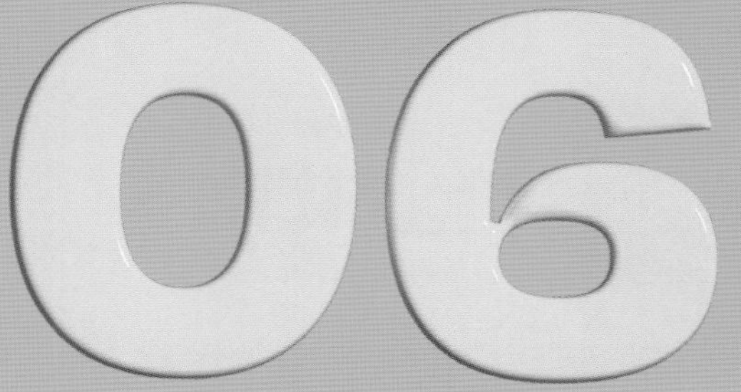

Have a drink, BTS.

<Run BTS!> Special Episode - Telepathy Part 1

It's been a while.

오랜만이에요.

o-raen-ma-ni-e-yo

As BTS starts filming <Run BTS!> and clinks glasses, j-hope says, "오랜만이에요 o-raen-ma-ni-e-yo," which means "It's been a while." He says this because it has been a while since they filmed <Run BTS!>. When you haven't seen someone or done something in a while, you can use this expression. If you go to a BTS concert after not having been to one in a long time, try saying "오랜만이야 o-raen-ma-ni-ya" or "오랜만이에요 o-raen-ma-ni-e-yo."

05
24
이상하다?
i-sang-ha-da
주워도 주워도 없어지지 않는 탁구공

Ping-pong balls that never seem to disappear no matter how many times he picks them up

<Run BTS!> Ep.139

That's weird.

이상하다?

i-sang-ha-da

While Jin is cleaning up after playing table tennis, ping-pong balls keep showing up no matter how hard he tries to put them all away. He says to himself, "**이상하다** i-sang-ha-da?" which means that something is weird. You can say this to express doubt about things that are unusual or strange. For example, if your meal seems to not get any smaller no matter how much you eat, or if your voice sounds different than usual because of a cold, try using this expression: **이상하다** i-sang-ha-da? **이상하네요** i-sang-ha-ne-yo?

Jung Kook | 반갑습니다 날씨가 많이 풀렸네요

Nice to meet you. The weather has warmed up a lot.

BTS (방탄소년단)
Jin's BE-hind 'Full' Story

The weather has warmed up a lot.

날씨가 많이 풀렸네요.

nal-sshi-ga ma-ni pul-lyeon-ne-yo

After a greeting, talking about the weather (날씨 nal-sshi) is the best way to continue a natural conversation. Jung Kook, who is meeting Jin after a long time apart, greets him and says, "날씨가 많이 풀렸네요 nal-sshi-ga ma-ni pul-lyeon-ne-yo." This expression is used when the weather is getting warmer, usually when winter is turning into spring. When a cold spell is ending, you can use this expression: 날씨가 많이 풀렸어 nal-sshi-ga ma-ni pul-lyeo-sseo, 날씨가 많이 풀렸네요 nal-sshi-ga ma-ni pul-lyeon-ne-yo.

23

j-hope: What did you order for the main dish?

<Run BTS!> Ep.130

What did you order?

뭐 시켰어?

mwo shi-kyeo-sseo

BTS is having a meal together! j-hope asks what the others ordered for their main dish by saying, "뭐 시켰어 mwo shi-kyeo-sseo?" In this expression, "시키다 shi-ki-da" means "to order something." Therefore, the expression asks what (뭐 mwo) has been ordered. If you would like to ask what someone ordered before you, you can say "뭐 시켰어 mwo shi-kyeo-sseo?" or "뭐 시켰어요 mwo shi-kyeo-sseo-yo?"

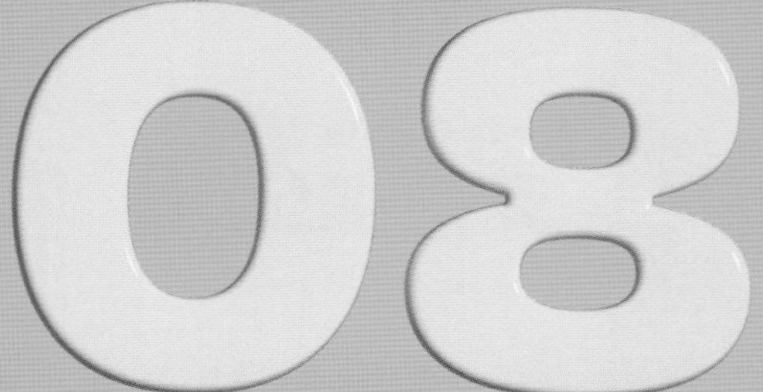

Tennis champion Jin is treating.

<Run BTS!> Ep.130

Thank you.

고마워요.

go-ma-wo-yo

RM says "고마워요 go-ma-wo-yo" to Jin, who is treating the others to lunch. As "감사합니다 gam-sa-ham-ni-da" has the same meaning, both expressions of gratitude can be used for everyone. You can also say "고마워 go-ma-wo" between close friends. Try using this expression anytime someone does something you are grateful for. 고마워 go-ma-wo! 고마워요 go-ma-wo-yo!

05

<Weverse Live> 2022.03.01

We do it sometimes.

가끔 해요.

ga-kkeum hae-yo

During a live video stream, SUGA is asked if BTS talks about the posts they upload on social media. He replies, "**가끔 해요** ga-kkeum hae-yo," which means "We do it sometimes." When someone asks how often you work out and you do it once or twice a week, try saying "**가끔 해** ga-kkeum hae," or "**가끔 해요** ga-kkeum hae-yo."

08

09

죄송합니다.
joe-song-ham-ni-da

카메라 추락사고

A camera-dropping accident

<Run BTS!> Ep.126

I apologize.

죄송합니다.

joe-song-ham-ni-da

Jin accidentally drops a camera and says "죄송합니다 joe-song-ham-ni-da," which is a very formal apology. Even though "미안합니다 mi-an-ham-ni-da" has the same meaning, it is better to use "죄송합니다 joe-song-ham-ni-da" when you feel strongly about your apology. You can also say "죄송해요 joe-song-hae-yo" in informal settings or "미안해 mi-an-hae" between close friends.

21

<Run BTS!> Ep.123

There is a problem.
문제가 생겼어.
mun-je-ga saeng-gyeo-sseo

After kneading some dough, Jin finds that it is sticking to the plastic bag. He says, "**문제가 생겼어** mun-je-ga saeng-gyeo-sseo," meaning "There is a problem." In this expression, "**문제** mun-je" means "problem," and "**생기다** saeng-gi-da" means "to occur." Therefore, it implies that a problem has arisen. When you encounter an unexpected issue while doing something, try saying "**문제가 생겼어** mun-je-ga saeng-gyeo-sseo" or "**문제가 생겼어요** mun-je-ga saeng-gyeo-sseo-yo."

Thank you for your effort.

수고하셨습니다.

su-go-ha-shyeot-seum-ni-da

After an interview or filming session, BTS usually says "수고하셨습니다 su-go-ha-shyeot-seum-ni-da!" This expression can be used to acknowledge people's hard work after they've finished a task. "수고 su-go" means "making an effort to do something." You can also say "수고했어 su-go-hae-sseo" to your close friends. At the end of the day, let's show our appreciation to the people around us. 수고했어 su-go-hae-sseo! 수고하셨습니다 su-go-ha-shyeot-seum-ni-da!

05

20

<Run BTS!> Ep.148

H: 하얗게 불태웠어요.
ha-ya-ke bul-tae-wo-sseo-yo

I'm burned out.

하얗게 불태웠다.

ha-ya-ke bul-tae-wot-da

RM finally succeeds in assembling a full-length mirror! At that moment, the subtitle "하얗게 불태웠다 ha-ya-ke bul-tae-wot-da" appears. This expression is used when someone has stayed up all night or worked hard for a specific task or job. It's like the work has burned them to ashes, just like trees turn to white ashes when they are burned. When you finally complete a demanding task, use the expression "하얗게 불태웠다 ha-ya-ke bul-tae-wot-da" or "하얗게 불태웠어요 ha-ya-ke bul-tae-wo-sseo-yo."

* SUGA did not attend due to personal reasons.

<Run BTS!> Ep.143

Hello?

여보세요?

yeo-bo-se-yo

BTS members call SUGA on the phone while filming <Run BTS!>. SUGA says "여보세요 yeo-bo-se-yo?" when answering the phone. This is how you say "Hello?" in Korean when you pick up the phone. When you answer the phone from a friend studying Korean with you, try this expression: 여보세요 yeo-bo-se-yo?

Only photos remain!
<Run BTS!> Ep.106

Only photos remain.
남는 건 사진뿐
nam-neun geon sa-jin-ppun

At the end of a photo exhibition in an episode of <Run BTS!>, BTS recommends that ARMY take pictures to capture good memories. j-hope says, "남는 건 사진뿐 nam-neun geon sa-jin-ppun," which literally means "Only photos remain." In this expression, "N (noun)뿐 ppun" means "There is only N (noun)," or "nothing but N (noun)." Do you agree with what j-hope says? If you go on a trip, you can certainly relate to this phrase.

I can do it!

<Run BTS!> Ep.123

H: 할 수 있어요.
hal ssu i-sseo-yo

I can do it.

hal ssu i-sseo

Jin is considering whether he should stop trying to make *kalguksu* (noodle soup), which is a food that is very hard to make. He encourages himself by shouting "할 수 있어 hal ssu i-sseo," which means "I can do it." When you feel like giving up during a demanding or challenging task, you can say this expression like Jin: 할 수 있어 hal ssu i-sseo! 할 수 있어요 hal ssu i-sseo-yo!

05

18

Do whatever you want with your free time today, BTS.

<Run BTS!> Ep.151

May 18th

H: 하고 싶은 거 다 하세요.
ha-go shi-peun geo da ha-se-yo

Do whatever you want.
하고 싶은 거 다 해.
ha-go shi-peun geo da hae

BTS is hanging out in the hotel during the staycation episode of <Run BTS!>. Everyone is enjoying their free time in a different way. The subtitles say, "방탄 하고 싶은 거 다 해 bang-tan ha-go shi-peun geo da hae." The expression "N (noun) 하고 싶은 거 다 해 ha-go shi-peun geo da hae" can be used when encouraging your close friends to do whatever they want. If you put their nickname in the "N (noun)" spot, it will sound more affectionate!

08

You're doing great.

[2020 FESTA] BTS (방탄소년단) '방탄생파'

August 13th

You're doing great.

잘하고 있어요.

jal-ha-go i-sseo-yo

SUGA, who is not a confident baker, is full of doubt as he tries to make a cake. Jimin encourages him by saying "잘하고 있어요 jal-ha-go i-sseo-yo." This expression is used to encourage someone who is struggling. When someone lacks confidence or is gripped by self-doubt, pat them on the back and say: 잘하고 있어 jal-ha-go i-sseo, 잘하고 있어요 jal-ha-go i-sseo-yo.

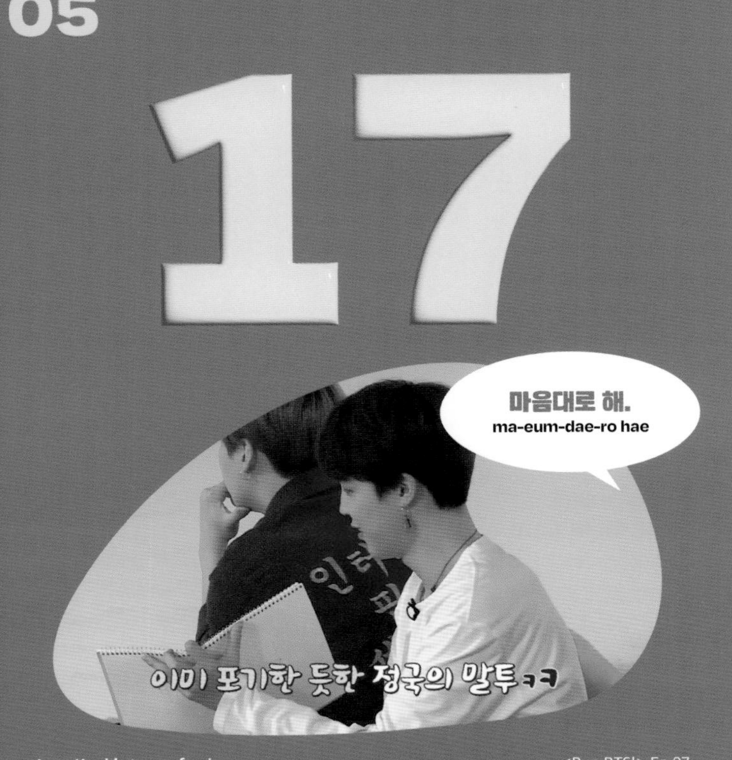

05
17
마음대로 해.
ma-eum-dae-ro hae
이미 포기한 듯한 정국의 말투ㅋㅋ

Jung Kook's tone of voice
that seems to have given up already lol

<Run BTS!> Ep.97

H: 마음대로 하세요.
ma-eum-dae-ro ha-se-yo

Do as you please.

마음대로 해.

ma-eum-dae-ro hae

BTS is playing a telepathy game. When it becomes hard to guess what the others are thinking, Jung Kook tells Jimin, "마음대로 해 ma-eum-dae-ro hae," which means "Do as you please." When you really don't care what you and your friends choose for lunch because you don't have much of an appetite, you can say "마음대로 해 ma-eum-dae-ro hae" or "마음대로 하세요 ma-eum-dae-ro ha-se-yo." However, keep in mind that this expression may sound a bit insincere.

08

Go for it, friends!

<Run BTS!> Special Episode
- Telepathy Part 1

August 14th

Go for it!

파이팅!

pa-i-ting

BTS is on a mission where everyone has to come up with the same place and gather there after a keyword is given. Jin shouts "파이팅 pa-i-ting!" to encourage the team in hopes of success. Derived from the English word "fighting," "파이팅pa-i-ting" is used when cheering for someone. It is sometimes pronounced as "화이팅 hwa-i-ting" as well. We wish you the best in learning Korean today. 파이팅 pa-i-ting!

16

There is hope.

<Run BTS!> Special Episode
- Next Top Genius Part 2

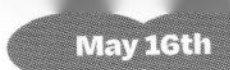

H: 희망이 보여요.
hi-mang-i bo-yeo-yo

There is hope.

희망이 보인다.

hi-mang-i bo-in-da

Jung Kook is pleased to have moved from the bottom of the rankings to the middle in a dice game. The subtitles describe this situation as "희망이 보인다 hi-mang-i bo-in-da," which means "There is hope." You can use this expression to express hope (희망 hi-mang) or faith in a promising future. When j-hope introduces himself, he says, "I'm j-hope, your 희망 hi-mang." When you see a silver lining after struggling with difficult problems, you can say "희망이 보인다 hi-mang-i bo-in-da!" or "희망이 보여요 hi-mang-i bo-yeo-yo!"

08

15

힘내!
him-nae

힘내라 우리 형

I'm cheering for you, hyung.

<Run BTS!> Ep.127

August 15th

H: 힘내세요!
him-nae-se-yo

I'm cheering for you!

him-nae

SUGA must get to a goal while hula hooping! Next to SUGA, RM roots for him by shouting, "힘내 him-nae!" This expression is used when cheering or showing support for someone to help them overcome a difficult situation. In honorifics, "힘내세요 him-nae-se-yo" can be said to make it sound formal and polite. Are you studying Korean today? 힘내세요 him-nae-se-yo!

05

15

힌트 투척
둘 중 하나야!

<Run BTS!> Ep.97

Hint dropping /

one of the two

둘 중 하나

dul jjung ha-na

During a game, Jung Kook has to draw a picture to indicate Jimin's word, which is based on j-hope's drawing representing "김태형 gim-tae-hyeong" (V's real name). j-hope and V give him a hint by saying the answer is "둘 중 하나 dul jjung ha-na," which means "one of the two." They are indicating that it is either "V" or "김태형 gim-tae-hyeong." This expression can be used when you have to choose between two options. When there are two menu options for an in-flight meal, you can say that you choose "둘 중 하나 dul jjung ha-na."

This was somehow very fulfilling.

BTS (방탄소년단) V's BE-hind 'Full' Story

It was fulfilling.
뿌듯하더라고요.
ppu-deu-ta-deo-ra-go-yo

V talks about writing the lyrics for <Blue & Grey> in an interview. Sharing his experience of using his computer to write lyrics instead of playing games, he uses the expression "뿌듯하더라고요 ppu-deu-ta-deo-ra-go-yo." "뿌듯하다 ppu-deu-ta-da" is an expression used when you are proud of yourself for accomplishing something. If you have reached a goal today, try using the expression "뿌듯해 ppu-deu-tae" or "뿌듯해요 ppu-deu-tae-yo."

14

This game is ridiculous.

<Run BTS!> Special Episode
- Next Top Genius Part 1

H: 어이가 없네요.
eo-i-ga eom-ne-yo

That's ridiculous.

어이가 없네.

eo-i-ga eom-ne

BTS is playing a card game! Unless Jung Kook has a card with the number "0" on it, he will lose the game. As he has had a bad hand from the start, he exclaims, "**어이가 없네** eo-i-ga eom-ne." The expression "**어이(가) 없다** eo-i(ga) eop-da" refers to a state in which you are left speechless because something unexpected has happened. For example, if someone responds angrily instead of apologizing for a mistake, you can say "**어이(가) 없네** eo-i(ga) eom-ne" or "**어이(가) 없네요** eo-i(ga) eom-ne-yo."

OK. I'll try again.

<Run BTS!> Special Episode - 'RUN BTS TV' On-air Part 1

I'll try again.

다시 해보겠습니다.

da-shi hae-bo-get-seum-ni-da

Jin is playing a game where he has to feed cakes to a bear. Even though he tries different ways to win the game, it doesn't work. Jin says, "다시 해보겠습니다 da-shi hae-bo-get-seum-ni-da," meaning "I'll try again." In this expression, "-겠습니다 get-seum-ni-da" is a polite expression to show a determination to do something. You already learned the phrase "잘 먹겠습니다 jal meok-get-seum-ni-da" (☛ July 21st), which means that you will enjoy the food. As you can see, this also includes "-겠습니다 get-seum-ni-da."

05

13

It worked.

<Run BTS!> Special Episode
- Mini Field Day Part 1

A: 통했어요.
tong-hae-sseo-yo

It worked.

통했다.

tong-haet-da

RM and V are playing one-on-one volleyball! RM scores with a light touch instead of a strong one. The subtitle describes it as "통했다 tong-haet-da," which means that a strategy or method was effective. If you have been studying Korean on your own using <365 BTS DAYS> and found it to be working well, you can say "통했다 tong-haet-da!" or "통했어요 tong-hae-sseo-yo!"

08

18

BTS' small happiness

<Run BTS!> Ep.151

a small happiness

소소한 행복

so-so-han haeng-bok

BTS finishes shooting a hotel staycation episode for <Run BTS!>. V thanks the production staff for giving them "소소한 행복 so-so-han haeng-bok." As "소소하다 so-so-ha-da" means "to be small," "소소한 행복 so-so-han haeng-bok" means "small happiness." When you are busy, you might overlook small happiness in your daily life. So try to find your own "소소한 행복 so-so-han haeng-bok" today!

05

12

<Run BTS!> Ep.136

Not confident /
I'll try my best!

H: 최선을 다해 볼게요.
choe-seo-neul da-hae bol-kke-yo

I'll try my best.

최선을 다해 볼게.

choe-seo-neul da-hae bol-kke

j-hope teams up with Jin for a quiz! Before they begin, j-hope says, "최선을 다해 볼게 choe-seo-neul da-hae bol-kke." A conjugation of "최선을 다하다 choe-seo-neul da-ha-da" (to do one's best), this expression is used when you promise to try your best. When talking about important exams or presentations, you can say "최선을 다해 볼게 choe-seo-neul da-hae bol-kke!" or "최선을 다해 볼게요 choe-seo-neul da-hae bol-kke-yo!"

08

19

pound (heart)

두근거리다

du-geun-geo-ri-da

During a live video stream, Jimin recalls his emotions on stage and says, "두근거리고 떨리고 du-geun-geo-ri-go tteol-li-go..." which means he felt his heart beating fast and trembling. "두근거리다 du-geun-geo-ri-da" means a state where the heart beats fast and hard. There is also an expression "두근두근 du-geun-du-geun" which can be used in almost all situations where the heart beats fast, such as excitement, surprise, and tension. To express your excitement before a BTS concert is about to begin, try saying "두근거려 du-geun-geo-ryeo" or "두근거려요 du-geun-geo-ryeo-yo."

11

H: 어쩔 수 없어요.
eo-jjeol ssu eop-seo-yo

There's nothing that can be done.

어쩔 수 없어.

eo-jjeol ssu eop-seo

BTS is singing old songs in a game in an episode of <Run BTS!>. As Jin's voice is a bit quiet, he can't achieve the score he wants. Then RM says, "어쩔 수 없어 eo-jjeol ssu eop-seo." This expression means "There's nothing that can be done." When you have important tasks to do but can't stop watching <Run BTS!> because it's hilarious, you can say "어쩔 수 없어 eo-jjeol ssu eop-seo," or "어쩔 수 없어요 eo-jjeol ssu eop-seo-yo."

08

20

cringing

이불킥

i-bul-kik

There are times when you lie down to sleep and cringe about things you said or did in the past. At that time, you may kick your blanket because you can't bring yourself to scream. "**이불킥** i-bul-kik" is a combination of "**이불** i-bul" (blanket) and "**킥** kik," the Korean spelling of the English word "kick." One of BTS songs, <Embarrassed>, describes the feeling of "**이불킥** i-bul-kik." In this song, the main character recalls the time when he acted awkwardly in front of a girl he likes.

05

10

<Run BTS!> Ep.124

It came to my mind.
생각이 났습니다.
saeng-ga-gi nat-seum-ni-da

Jimin is giving a presentation of new game ideas for <Run BTS!>. When the games he used to play as a child flash across his mind, he says, "생각이 났습니다 saeng-ga-gi nat-seum-ni-da." This is a conjugation of "생각(이) 나다 saeng-gak(i) na-da," which means "to come into one's mind." When recalling a Korean expression that you studied before but forgot, you can say "생각(이) 났어 saeng-gak(i) na-sseo!" or "생각(이) 났습니다 saeng-gak(i) nat-seum-ni-da!"

<Weverse Live> 2021.11.28

It's an honor.

영광입니다.

yeong-gwang-im-ni-da

After receiving a compliment from Jimin during a live video steram, V expresses his humility by saying "영광입니다 yeong-gwang-im-ni-da." When you have accomplished something great or feel very grateful, you can use this expression to show your humility. Try saying "영광입니다 yeong-gwang-im-ni-da" when you receive praise for something great you have achieved!

05

09

Hey, Jung Kook knew it at a glance.

<Run BTS!> Special Episode - Telepathy Part 1

May 9th

H: 보자마자 알았어요.
bo-ja-ma-ja a-ra-sseo-yo

He knew it at a glance.

보자마자 알았어.

bo-ja-ma-ja a-ra-sseo

BTS is taking a quiz where they have to guess words after reading their definitions, and Jung Kook instantly came up with the correct answer! RM says, "보자마자 알았어 bo-ja-ma-ja a-ra-sseo," which implies that Jung Kook knew the answer as soon as he saw the definition. You can use the expression "-자마자 ja-ma-ja" when you can do one action immediately after doing another. For example, if you can recognize BTS right away, even from a distance, you can say "보자마자 알았어 bo-ja-ma-ja a-ra-sseo," or "보자마자 알았어요 bo-ja-ma-ja a-ra-sseo-yo."

08

22

August 22nd

I choked up.

울컥했어요.

ul-keo-kae-sseo-yo

During a live video stream celebrating his birthday, Jin reminisces about the surprise birthday party thrown by ARMY at a concert, saying "울컥했어요 ul-keo-kae-sseo-yo." "울컥하다 ul-keo-ka-da" is an expression used to describe a strong emotional response, often leading to a feeling of being choked up or on the verge of tears. You can use this expression when looking back on meeting BTS at a concert for the first time: 울컥했어 ul-keo-kae-sseo, 울컥했어요 ul-keo-kae-sseo-yo.

05

08

It suits you well.

<Run BTS!> Ep.145

H: 잘 어울려요.
jal eo-ul-lyeo-yo

It suits you well.

잘 어울린다.

jal eo-ul-lin-da

j-hope is amazed by V, who is dressed in a warrior costume for <Run BTS!>. He says, "**잘 어울린다** jal eo-ul-lin-da," because the outfit and beard suit V well. "**어울리다** eo-ul-li-da" means that something is in harmony with something else, and "**잘** jal" means "well." When a new album concept fits BTS' style well, you can say "**잘 어울린다** jal eo-ul-lin-da" or "**잘 어울려요** jal eo-ul-lyeo-yo."

Jimin loses his heart to a new trendy item.

<Run BTS!> Ep.151

lose one's heart

홀딱 반하다

hol-ttak ban-ha-da

Jimin shows great interest in a toy car with a camera! The subtitles say "홀딱 반한 지민 hol-ttak ban-han ji-min." The word "홀딱 hol-ttak" means "completely," and "반하다 ban-ha-da" means "to fall in love with something." So "홀딱 반하다 hol-ttak ban-ha-da" means that someone has completely fallen in love with something or somebody or lost their heart to them. When you watch a wonderful BTS performance that makes you feel starstruck, you can use the expression "홀딱 반했어 hol-ttak ban-hae-sseo!" or "홀딱 반했어요 hol-ttak ban-hae-sseo-yo!" to convey that you are completely smitten with their performance.

05

07

[VLOG] 진 | 낚시 어디까지 가봤니?!

May 7th

H: 물 건너갔어요.
mul geon-neo-ga-sseo-yo

It has crossed the water.

물 건너갔다.

mul geon-neo-gat-da

Jin shares his experience of fishing in the ocean with professional fishers! When he asked them if he would see sharks, they answered, "No." Then he thought, "물 건너갔다 mul geon-neo-gat-da." This expression literally means that something has crossed the water. In this case, "물 mul" (water) is a generic term for a large river or sea. If something sailed across the sea, wouldn't it be hard to chase it? Therefore, "물 건너갔다 mul geon-neo-gat-da" can be used to say that something is irreversible or out of reach. Fortunately, Jin was able to spot a shark later that day!

08

24

자랑스럽다.
ja-rang-seu-reop-da

자랑스러운 777 1등 민슈가

Proud 777 winner, SUGA

<Run BTS!> Ep.127

H: 자랑스러워요.
ja-rang-seu-reo-wo-yo

I'm proud.

자랑스럽다.

ja-rang-seu-reop-da

SUGA says he is proud of himself for completing some challenging missions and going home first. You can express this kind of feeling of pride as "자랑스럽다 ja-rang-seu-reop-da," like SUGA does. The word "자랑 ja-rang" means "someone or something worth being proud of." Doesn't this sound similar to "사랑 sa-rang," which means "love?" That is why you can find the word "자랑 ja-rang" and "사랑 sa-rang" in the lyrics of the BTS song <Trivia 承 : Love>. When you are proud of BTS for moving people with their beautiful lyrics, try saying "자랑스럽다 ja-rang-seu-reop-da," or "자랑스러워요 ja-rang-seu-reo-wo-yo."

05

06

[BTS VLOG] Jung Kook I CAMPING VLOG

May 6th

It turned 180 degrees.

180도 변했습니다.

baek-pal-ship-do byeon-haet-seum-ni-da

When the sunny weather suddenly turns rainy, Jung Kook says, "180도 변했습니다 baek-pal-ship-do byeon-haet-seum-ni-da." This literally means that something has turned 180 degrees. As a 180-degree change means you're headed in the exact opposite direction, the expression implies that the weather has completely changed. Is the concept of BTS' new album totally different from the previous ones? Then try saying "180도 변했어 baek-pal-ship-do byeon-hae-sseo," or "180도 변했습니다 baek-pal-ship-do byeon-haet-seum-ni-da."

A fully satisfied facial expression

<Run BTS!> Ep.146

satisfied

흐뭇하다

heu-mu-ta-da

V has an important clue that is critical for the mission during an episode of <Run BTS!>. The subtitles say "흐뭇해진 표정 heu-mu-tae-jin pyo-jeong," which means "a satisfied facial expression." This is a conjugation of "흐뭇하다 heu-mu-ta-da" (to be satisfied). Is there something that makes you feel pleased and content just by looking at it? For instance, if a relaxing BTS getaway, where they have fun and enjoy tasty meals, make you feel happy, try saying "흐뭇해 heu-mu-tae" or "흐뭇해요 heu-mu-tae-yo."

05

05

Jin.

'달려라 방탄'이 조금 애착이 가네요

I'm a bit attached to the song <Run BTS>.

BTS (방탄소년단) 77Q 77A Interview

be attached to something

애착이 가다

ae-cha-gi ga-da

When Jin is asked which song from the new release he is most attached to, he says <Run BTS>. "애착이 가다 ae-cha-gi ga-da" can be used when you especially love something. "애착 ae-chak" means that you are so drawn to something that you cherish and love it. Do you have a favorite BTS song that mesmerizes you? Then try saying "애착이 가 ae-cha-gi ga" or "애착이 가요 ae-cha-gi ga-yo" while talking about it!

08

26

August 26th

I missed you.

그리웠어요.

geu-ri-wo-sseo-yo

Recalling the feeling of being on the concert stage after a long time during a live video stream, Jin says "그리웠어요 geu-ri-wo-sseo-yo." This expression is used when you finally come face to face with a person or situation you have missed. You can also say "보고 싶었어요 bo-go shi-peo-sseo-yo" (☛ February 13th), which means "I missed you." If BTS starts a live video stream, leave a comment like this: 그리웠어요 geu-ri-wo-sseo-yo.

It's unusual!

<Run BTS!> Special Episode - 'RUN BTS TV' On-air Part 1

H: 심상치 않아요.
shim-sang-chi a-na-yo

It's unusual!

심상치 않아!

shim-sang-chi a-na

When BTS gathers to film <Run BTS!>, they look around the site, feeling a different atmosphere than usual. At that moment, Jimin shouts, "심상치 않아 shim-sang-chi a-na!" This expression is used when a mood or situation is unusual, and something extraordinary or suspicious is about to happen. When a video or image teaser is uploaded to the official BTS account, use this expression: 심상치 않아 shim-sang-chi a-na! 심상치 않아요 shim-sang-chi a-na-yo!

08

27

<Weverse Live> 2022.06.07

August 27th

I mean it.

진심입니다.

jin-shi-mim-ni-da

As Jung Kook is finishing his live video stream, he says he always thinks of ARMY and adds, "**진심입니다** jin-shi-mim-ni-da." In this expression, the word "**진심** jin-shim" consists of "**진** jin" (to be real) and "**심** shim" (heart). You can use this expression when telling the truth from the heart. The BTS song <The Truth Untold> also contains the word "**진심** jin-shim." We wish you good luck in studying Korean. **진심입니다** jin-shi-mim-ni-da!

<Run BTS!> Ep.96

May 3rd

It doesn't matter.
상관없어요.
sang-gwan-eop-seo-yo

V and Jung Kook are about to play a classic Korean game called *ttakji*. Jin asks Jung Kook which piece he is going to use, and Jung Kook replies, "상관없어요 sang-gwan-eop-seo-yo." This expression means that it doesn't matter which one you choose. If it doesn't matter whether you have pizza or chicken for dinner, try using the expression "상관없어 sang-gwan-eop-seo" or "상관없어요 sang-gwan-eop-seo-yo."

08

28

Shy

<Run BTS!> Ep.99

H: 쑥스럽네요.
ssuk-seu-reom-ne-yo

I'm flattered.

쑥스럽구만.

ssuk-seu-reop-gu-man

Jimin has made a beautiful flower arrangement! Feeling shy because of the instructor's compliment for his delicate work, he says, "쑥스럽구만 ssuk-seu-reop-gu-man." "쑥스럽다 ssuk-seu-reop-da" refers to a state in which someone is so bashful that they are acting in an awkward or embarrassed manner. You can use it when you are praised like Jimin or feel awkward in front of strangers or someone you like. You can say "쑥스럽네요 ssuk-seu-reom-ne-yo" to make it sound formal and polite. If someone compliments your Korean skills, try using the expression "쑥스럽구만 ssuk-seu-reop-gu-man" or "쑥스럽네요 ssuk-seu-reom-ne-yo."

05

j-hope | 어쨌거나 저는 키워드를 되게 중요하게 생각해요

Anyway, I consider keywords to be very important.

BTS (방탄소년단)
j-hope's BE-hind 'Full' Story

I find it important.

중요하게 생각해요.

jung-yo-ha-ge saeng-ga-kae-yo

V asks j-hope what is important when creating music, and j-hope answers that keywords matter. When you think something is important, you can say "N (noun)을/를 중요하게 생각해요 eul/reul jung-yo-ha-ge saeng-ga-kae-yo." What do you think is essential in life? Use the expression "중요하게 생각해 jung-yo-ha-ge saeng-ga-kae" or "중요하게 생각해요 jung-yo-ha-ge saeng-ga-kae-yo." You can add what you think is important in front of this phrase.

* Nouns ending with a batchim use "을 eul," and nouns ending without one use "를 reul."

08

29

뜻깊은 시간
tteut-gi-peun shi-gan

잘 참고해서
좋은 무대로 보답하겠습니다

We'll take it into consideration and repay with a great performance.

<Run BTS!> Ep.144

a meaningful time
뜻깊은 시간
tteut-gi-peun shi-gan

BTS has checked out the results of a survey of ARMY about their songs! RM says it was meaningful because he now has a better understanding of ARMY's opinions and will reflect them in future performances. He uses the expression "뜻깊은 시간 tteut-gi-peun shi-gan." As "뜻깊다 tteut-gip-da" literally means "to be deeply meaningful," the phrase "뜻깊은 시간 tteut-gi-peun shi-gan" implies that the shared time is valuable and significant. Every moment BTS and ARMY have shared is "뜻깊은 시간 tteut-gi-peun shi-gan" too!

05

01

May 1st

H: 운이 좋아요.
u-ni jo-a-yo

He's lucky.

운이 좋다.

u-ni jo-ta

BTS is playing a dice game in an episode of <Run BTS!>, and SUGA keeps taking the lead with good rolls. Watching him, Jung Kook says, "운이 좋다 u-ni jo-ta." As "운 un" means "force of luck" and "좋다 jo-ta" means "to be good," "운이 좋다 u-ni jo-ta" implies that things are working out fine due to good luck. On the contrary, when you have bad luck, you can say, "운이 나쁘다 u-ni na-ppeu-da." If you consider someone lucky, you can use the expression "운이 좋다 u-ni jo-ta" or "운이 좋아요 u-ni jo-a-yo."

08

<Weverse Live> 2021.02.27

a meaningful day

의미 있는 하루

ui-mi in-neun ha-ru

When Jung Kook wraps up his live video stream, he says he hopes ARMY had "의미 있는 하루 ui-mi in-neun ha-ru" today. This expression consists of "의미 있다 ui-mi it-da" (to be meaningful) and "하루 ha-ru" (a day). We can feel Jung Kook's love for ARMY as he hopes that ARMY has had another meaningful and fulfilling day. Similarly, today is another day of learning Korean with BTS. Was it "의미 있는 하루 ui-mi in-neun ha-ru" for you too?

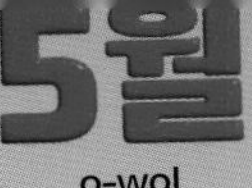

May

August 31st

precious and good emotions

소중하고 좋은 감정

so-jung-ha-go jo-eun gam-jeong

When Jin is asked what the most precious emotion (감정 gam-jeong) is, he answers that happiness, fun, joy, and amusement are "소중하고 좋은 감정 so-jung-ha-go jo-eun gam-jeong." "소중하다 so-jung-ha-da" means "to be precious" and "좋다 jo-ta" means "to be good." You don't have to use the two words simultaneously—you can put a different modifier in front of "감정 gam-jeong." The excitement that ARMY feels when watching or thinking of BTS can be called "소중하고 좋은 감정 so-jung-ha-go jo-eun gam-jeong," right?

04

30

<Run BTS!> Ep.132

In *sikhye*, the liquid and the rice grains are

one and the same

마찬가지

ma-chan-ga-ji

The traditional Korean beverage *sikhye* (sweet rice punch) contains whole rice grains in it. BTS is discussing whether they eat the rice grains or not. Jin argues that the drink and rice grains are two parts of a single entity, just like the seven members of BTS. He uses the expression "**마찬가지** ma-chan-ga-ji," which means that something, such as a status, shape, or opinion, is identical to another. If someone tells you that they love BTS new song, you can respond by saying "**마찬가지야** ma-chan-ga-ji-ya!" or "**마찬가지예요** ma-chan-ga-ji-ye-yo!" to express that you feel the same way.

September

04

first and last

처음이자 마지막

cheo-eu-mi-ja ma-ji-mak

After his interview with V, j-hope remarks that it is an unusual occurrence and adds, "**처음이자 마지막** cheo-eu-mi-ja ma-ji-mak," which means that this is both the first (**처음** cheo-eum) and last (**마지막** ma-ji-mak) time. In other words, he is indicating that this is a rare or unique event that is unlikely to be repeated. For example, if BTS is the group that makes you an avid fan for the first time, and you believe you will not become an avid fan of any other group like BTS, you can say "**처음이자 마지막** cheo-eu-mi-ja ma-ji-mak" to express that BTS is the first and last group that you will support in this way.

01

Happy Birthday, Jung Kook!

생일 축하해요, 정국!

saeng-il chu-ka-hae-yo, Jung Kook

04

Scratching

<Run BTS!> Ep.111

scratching

geuk-jeok

While adding up some numbers in a game, SUGA scratches his head in surprise because of an unexpected result. The word "긁적 geuk-jeok" appears on the screen, which refers to the gesture of scratching. We often scratch our head or other body parts during embarrassing or awkward moments. In that case, use the phrase "긁적 geuk-jeok" repeatedly as "긁적긁적 geuk-jeok-geuk-jeok" to express that you are continuously scratching something.

Excitedly shaking

<Run BTS!> Ep.152

excitedly shaking

deul-sseok-deul-sseok

In a guessing game of old songs, when the bridge of a song plays, Jimin dances with excitement, shaking his shoulders and hips. The subtitles at that moment read "들썩들썩 deul-sseok-deul-sseok," an expression used to describe people shaking their shoulders or hips. Have you ever found yourself moving to the rhythm of BTS' mind-blowing songs? If so, try using the expression "들썩들썩 deul-sseok-deul-sseok" to describe the feeling of excitement and movement.

04

Teary eyes

<Run BTS!> Special Episode
- Next Top Genius Part 1

teary eyes

그렁그렁

geu-reong-geu-reong

While the others are enjoying the pizza, V, who needs to go on a diet, is just looking at it. With a desire to eat together, V's eyes become moist with tears, and the subtitle "그렁그렁 geu-reong-geu-reong" appears on the screen. This expression represents the appearance of tears welling up in the eyes. Imagine you ever get to meet BTS in person. Don't you think you might become overwhelmed and teary-eyed, just like "그렁그렁 geu-reong-geu-reong?"

09

A throbbing heart

<Run BTS!> Ep.140

a throbbing heart
콩닥콩닥
kong-dak-kong-dak

BTS is in a state of tension before a game in which they have to give up food one at a time as a penalty if they can't guess the identity of the person in a picture. Just then, the subtitle "콩닥콩닥 kong-dak-kong-dak" appears. This refers to the sound or feeling of a pounding heart. A synonym for this is "두근두근 du-geun-du-geun" (👉 August 19th). How would you feel if you met BTS and had a conversation with them? Your heart might thump with excitement! 콩닥콩닥 kong-dak-kong-dak!

04

26

April 26th

blurry

가물가물

ga-mul-ga-mul

BTS is taking a quiz in an episode of <Run BTS!>. The answer to a question is "sepak takraw," a sport similar to foot volleyball, but SUGA is frustrated by his inability to remember it. The expression "**가물가물** ga-mul-ga-mul" can be used in this situation. It is used to describe a vague and uncertain memory, like a faint image that seems to disappear, or a faint movement. So, you can use it when you have a vague recollection and can't remember things well. If someone asks you about something that happened when you were young and you don't remember it well, try using the expression "**가물가물해** ga-mul-ga-mul-hae," or "**가물가물해요** ga-mul-ga-mul-hae-yo."

09

04

둠칫

둠칫

Boom-bop-boom-bop

<Run BTS!> Ep.134

boom-bop-boom-bop

둠칫둠칫

dum-chit-dum-chit

BTS plays a game where they have to guess the title of a song by reading a description of its choreography. V is moving to the choreography being described, and the subtitle "**둠칫둠칫** dum-chit-dum-chit" appears next to him. This expression is originally used to represent rhythmic drum sounds, but nowadays it refers to the movement of dancing to the beat. When you find yourself moving along to the rhythm of a BTS song, try using the expression "**둠칫둠칫** dum-chit-dum-chit."

04

Congratulations on the 100th episode special of <Run BTS!> / Quickly

<Run BTS!> Ep.100

quickly

hu-da-dak

BTS is drawing lots to choose an item to replace their badminton rackets! When V moves faster than the others to draw lots, the subtitle "후다닥 hu-da-dak" appears next to him. This expression refers to the shape when you quickly move your body or hustle with your work. If you rush through your studies or work to watch <Run BTS!>, you can use the expression "후다닥 hu-da-dak."

A shy laugh

<Run BTS!> Ep.146

a shy laugh

헤헷

he-het

When Jung Kook fortunately finds all the hidden objects on <Run BTS!>, the subtitle "**헤헷** he-het" appears next to him. This expression represents the sound of a shy laugh. When someone praises you, or you accomplish something, you can express your feelings of happiness and bashfulness by saying "**헤헷** he-het." Similarly, you can say "**헤헤** he-he." Do BTS' adorable pictures make you smile? Then use the expression "**헤헷** he-het!"

04

What kind of serve was that? lol

* SUGA did not attend due to personal reasons.
<Run BTS!> Ep.139

April 24th

tiptoe
살금살금
sal-geum-sal-geum

j-hope and V are playing table tennis! Despite his excellent form, j-hope's serve bounces weakly over the net. RM says, "살금살금 sal-geum-sal-geum," to describe the ball's slow movement, comparing it to someone walking on tiptoe. This is an expression used when moving carefully so that nobody will notice you. If you are in a quiet place like a library, you must walk on your toes to remain silent. 살금살금 sal-geum-sal-geum!

Memories recalled

<Run BTS!> Ep.144

memories recalled

추억이 방울방울

chu-eo-gi bang-ul-bang-ul

BTS is reminiscing about their best performances as chosen by ARMY, and the subtitle "**추억이 방울방울** chu-eo-gi bang-ul-bang-ul" appears. "**추억** chu-eok" means "memory," and "**방울** bang-ul" means "droplet" or "bubble." "**방울방울** bang-ul-bang-ul" refers to multiple droplets or bubbles gathered together. Thus, "**추억이 방울방울** chu-eo-gi bang-ul-bang-ul" describes a state in which lots of old memories come to mind, much like beautifully rising soap bubbles. When you watch old BTS videos, you might experience this feeling of "**추억이 방울방울** chu-eo-gi bang-ul-bang-ul."

04

Nervous that he might get caught

<Run BTS!> Ep.140

nervous
조마조마
jo-ma-jo-ma

In a game where BTS must strike the same pose, Jin, RM, and j-hope all make mistakes! Jin and RM's mistakes have been caught, but j-hope waits nervously to see if anybody will notice his. The subtitles describe him as "**조마조마** jo-ma-jo-ma." When you are anxious or worried about upcoming events, you can use this expression. If you are worried a minute before booking BTS concert tickets, try using the expression: **조마조마해** jo-ma-jo-ma-hae! **조마조마해요** jo-ma-jo-ma-hae-yo!

09

Everyone is picking up clues one by one to prove their innocence.

<Run BTS!> Ep.145

September 7th

picking up one by one

주섬주섬

ju-seom-ju-seom

BTS must find the member who hid some clues. They decide to share the information they have gathered. Everyone starts to take out clues from their pockets, pouches, and nearby places, and "주섬주섬 ju-seom-ju-seom" appears in the subtitles to describe their actions. This expression is commonly used when picking up scattered items one by one. You can also use this expression when picking up scattered coins one by one from the floor.

04

Perked-up ears

<Run BTS!> Ep.153

perked-up ears

쫑긋

jjong-geut

BTS is currently practicing an old song by dividing it into parts. When Jung Kook starts singing SUGA's part, SUGA's ears perk up. The subtitle describes the situation as "쫑긋 jjong-geut," which refers to the shape of an animal's ears when they stand erect to better hear a sound. If you overhear someone talking about BTS, you might strain your ears to hear exactly what they're saying. 쫑긋 jjong-geut!

Blazing

<Run BTS!> Ep.96

blazing

이글이글

i-geu-ri-geul

Before a spinning top match, Jimin's eyes burn with determination to win as he glares at the camera. At that moment, the subtitle "이글이글 i-geu-ri-geul" appears along with fire-shaped graphics. This expression describes the image of leaping flames with rising heat. It represents the intense emotions of passion and the burning desire to win, like a fire that cannot be extinguished. We hope you continue to study Korean with the same fiery passion! 이글이글 i-geu-ri-geul!

04

<Run BTS!> Ep.103

Speaking clearly /

speaking clearly
또박또박
tto-bak-tto-bak

RM and Jimin are cooking by following SUGA's directions! Clearly reiterating SUGA's instructions, RM makes Jimin put in the meat first. At that moment, the subtitle "또박또박 tto-bak-tto-bak" shows up. This is an expression that represents a clear and articulate way of speaking. Let's study Korean by saying each word. 또박또박 tto-bak-tto-bak!

V's birthday

<Weverse Live> 2020.12.30

September 9th

in a hurry

부랴부랴

bu-rya-bu-rya

In a live video stream, V says he came in a hurry to meet ARMY, using the expression "부랴부랴 bu-rya-bu-rya," which means "doing something in a hurry." When a fire (불bul) breaks out, Koreans say, "불이야 bu-ri-ya! 불이야 bu-ri-ya!" to warn others. If you pronounce it quickly, it sounds like "부랴부랴 bu-rya-bu-rya." That's why we use this expression to mean "in haste like something is on fire."

20

<Run BTS!> Special Episode - Telepathy Part 2

yummy

nyam-nyam

Jung Kook is enjoying some ice cream! At that moment, the expression "냠냠 nyam-nyam" appears on the screen. This refers to the sound or act of someone enjoying food. When you have a mouthful of delicious food and are chewing on it, you can say, "냠냠 nyam-nyam." You can also use this expression when describing BTS eating something tasty.

10

Boohoo

<Run BTS!> Ep.128

boohoo

heu-keuk

During the Liar Game, Jin fails to guess the keyword that only he doesn't know. At that moment, the subtitle "흑흑 heu-keuk" appears next to his face. This expression refers to the sound of sobbing. "엉엉 eong-eong" is another expression used to describe the sound of unrestrained crying. When you feel emotional because you miss BTS, try saying "흑흑 heu-keuk" to express your feelings.

04

gulping

벌컥벌컥

beol-keok-beol-keok

While playing the Liar Game during an episode of <Run BTS!>, Jin doesn't know the keyword, but he must make up explanations as if he does. When he gulps down a glass of water, "벌컥벌컥 beol-keok-beol-keok" appears on the screen. This is an expression that describes the sound of drinking something. When someone takes a drink of water after jumping up and down during a BTS concert, you can describe it by saying "벌컥벌컥 beol-keok-beol-keok."

* SUGA did not attend due to personal reasons.
<Run BTS!> Ep.139

lightly

살살

sal-sal

During a table tennis match with Jin, j-hope hits the ball too hard and keeps conceding points. RM advises him to relax and swing less forcefully by saying "살살 sal-sal," which means "doing something softly and lightly." If someone is walking heavily while a baby is sleeping, you can use the expression "살살 sal-sal" to ask them to walk more softly and gently.

04

OK. Call.

<Run BTS!> Ep.149

kol

Jimin, who wants to move his desk, asks Jung Kook for help. Although Jung Kook is doing a task, he agrees without hesitation and helps. At that moment, the subtitles say "**OK 콜** kol." "콜 kol" is the Korean spelling of "call" in English and is used when agreeing to a request or accepting a challenge. Should we accept a friend's invitation to study Korean together while watching BTS video clips? 콜 kol!

12

Happy Birthday, RM!

생일 축하해요, 알엠!

saeng-il chu-ka-hae-yo, RM

The Bear team attempts a twist but fails!

<Run BTS!> Ep.113

April 17th

a twist

반전

ban-jeon

BTS pairs up to play a version of limbo. Each person must carry another with their backs touching and pass under the limbo stick! The lower the stick, the better it is for a smaller person to carry a bigger one. However, when V, who is relatively big, carries SUGA, Jin says, "**반전이네** ban-jeo-ni-ne." When you encounter a situation opposite from what is expected, you can call it "**반전** ban-jeon." When BTS' newly released songs have different vibes from their teaser, you can say, "**반전이네** ban-jeo-ni-ne!" or "**반전이네요** ban-jeo-ni-ne-yo!"

Don't worry.

BTS (방탄소년단) 77Q 77A Interview

Don't worry.

걱정하지 말아요.

geok-jeong-ha-ji ma-ra-yo

Jimin tells ARMY not to worry by saying "**걱정하지 말아요** geok-jeong-ha-ji ma-ra-yo." How sweet! In this expression, "**-지 말아요** ji ma-ra-yo" means "not to do something." You can casually say "**걱정하지 마** geok-jeong-ha-ji ma" to your close friends. When someone is very worried, try saying this expression to reassure them: **걱정하지 마** geok-jeong-ha-ji ma, **걱정하지 말아요** geok-jeong-ha-ji ma-ra-yo.

04

j-hope's inner conflict

<Run BTS!> Ep.144

inner conflict
내적 갈등
nae-jjeok gal-tteung

When asked to guess their top summer song chosen by ARMY, j-hope can't decide between <So What> and <Dynamite>. The subtitles describe this as "**내적 갈등** nae-jjeok gal-tteung." This refers to an inner psychological conflict. You may feel this when you can't choose one thing from many options. If you can't decide on your favorite BTS summer song, like j-hope, try using the expression "**내적 갈등** nae-jjeok gal-tteung."

I'm rooting for you.

당신을 응원합니다.

dang-shi-neul eung-won-ham-ni-da

Jin has temporarily become a content creator during a live video stream. After everyone else has left, SUGA stays behind and says "**당신을 응원합니다** dang-shi-neul eung-won-ham-ni-da" to show his support for Jin. You can casually say "**너를 응원해** neo-reul eung-won-hae" to your close friends. Is there anyone who needs your support and encouragement? Then try saying "**너를 응원해** neo-reul eung-won-hae" or "**당신을 응원합니다** dang-shi-neul eung-won-ham-ni-da."

I'm having a mental breakdown... Let's go...

<Run BTS!> Ep.153

mental breakdown

멘붕

men-bung

BTS has to divide up an old song, memorize each part, and sing it to the end! When V finds this task more challenging than expected, he says, "**멘붕** men-bung." This is a new word that combines "**멘탈** men-tal," the Korean spelling of the English word "mental," and "**붕괴** bung-goe," which means "collapse." You can use this word when you feel overwhelmed or flustered. For instance, when you realize you forgot to bring your wallet or phone with you, you can exclaim, "**멘붕** men-bung!"

15

세상 따숩
틀려도 괜찮아~ 그럴 수도 있지!

The warmest person in the world /
It's okay to be wrong. It could happen!

<Run BTS!> Ep.153

H: 틀려도 괜찮아요.
teul-lyeo-do gwaen-cha-na-yo

It's okay to be wrong.

틀려도 괜찮아.

teul-lyeo-do gwaen-cha-na

In a mission to sing a line from a song correctly, V gets disappointed when he makes a mistake. To cheer him up, Jimin says "**틀려도 괜찮아** teul-lyeo-do gwaen-cha-na," which is the conjugated form of "**틀리다** teul-li-da" (to be wrong). You can use the expression "**-아/어도 괜찮아** a/eo-do gwaen-cha-na" to reassure someone that it's okay. For instance, if somebody is worried about being late for an appointment, you can say "**늦어도 괜찮아(요)** neu-jeo-do gwaen-cha-na(yo)," conjugating "**늦다** neut-da" (to be late).

04

14

황당

Baffled

<Run BTS!> Ep.112

baffled

hwang-dang

When V speaks in a funny accent, Jin stares at him with a baffled look. At that moment, the expression "황당 hwang-dang" appears on the screen. This comes from the basic form "황당하다 hwang-dang-ha-da," which means "to be baffled by someone's words or actions." It can be used when someone is puzzled by something that another person has said or done. For instance, if you have been waiting for the bus for 30 minutes and it passes your stop, you might be at a loss for words. In such a situation, you could text your friend "황당 hwang-dang" with a shocked emoji!

catch on quickly

눈치가 빠르다

nun-chi-ga ppa-reu-da

j-hope quickly approaches Jin, who discovered the clue of a game. At that moment, Jin describes j-hope as "**눈치가 빠르다** nun-chi-ga ppa-reu-da." "**눈치** nun-chi" means "the ability to read the room or the people's minds," and "**빠르다** ppa-reu-da" means "to be fast." Therefore, "**눈치가 빠르다** nun-chi-ga ppa-reu-da" is used to describe people who quickly catch on to the atmosphere or a person's mood. If you are planning a surprise birthday party for your friend and they find out, you can say, "**눈치가 빨라** nun-chi-ga ppal-la" or "**눈치가 빨라요** nun-chi-ga ppal-la-yo."

<Run BTS!> Ep.145

13

<Run BTS!> Ep.155

quivering eyes
동공 지진
dong-gong ji-jin

As Jung Kook is about to answer a question in a quiz, he is startled to see several different types of water guns that will be used as a penalty for giving the wrong answer. "동공 지진 dong-gong ji-jin" appears in the subtitles. This expression describes a person's pupils (동공 dong-gong) trembling in fear, as if there were an earthquake (지진 ji-jin). When a baffling situation causes your eyes to tremble, you can use the expression "동공 지진 dong-gong ji-jin."

09

17

머리가 하얘진다.
meo-ri-ga ha-yae-jin-da

Full of worries even before starting

<Run BTS!> Ep.96

September 17th

H: 머리가 하얘져요.
meo-ri-ga ha-yae-jeo-yo

I have gone blank.

머리가 하얘진다.

meo-ri-ga ha-yae-jin-da

While assembling a minicar, j-hope uses the expression "**머리가 하얘진다** meo-ri-ga ha-yae-jin-da," being concerned that it will take too long to build it. Was there a moment when you were so worried that you couldn't do anything else? "**머리가 하얘진다** meo-ri-ga ha-yae-jin-da" is used when you can't think because the inside of your head (**머리** meo-ri) has gone as blank as a white sheet of paper. Wouldn't your head go blank if you had the opportunity to speak with BTS?

<Run BTS!> Ep.153

ditto
동감
dong-gam

Jin agrees with the others that RM's singing performance is touching! The subtitle uses "**동감** dong-gam" to describe this situation. This expression is used when you have the same idea or feelings as someone else. If you relate to someone who says that BTS' new album is great, try saying "**동감이야** dong-ga-mi-ya" or "**동감이에요** dong-ga-mi-e-yo." Or you can simply say "**동감** dong-gam!" to your close friends.

18

<Run BTS!> Ep.122

tilt one's head

고개를 갸우뚱거리다

go-gae-reul gya-u-ttung-geo-ri-da

Jin tastes the sauce that he is making for *tangsuyuk* (sweet and sour pork). He tilts his head and puts on a clueless face after tasting it. When something is unclear or difficult to understand, you tilt your head to one side, which is described as "고개를 갸우뚱거리다 go-gae-reul gya-u-ttung-geo-ri-da" in Korean. Sometimes the head-tilting gesture is separately described as "갸우뚱 gya-u-ttung" in the subtitles.

Can't hold back laughter

<Run BTS!> Ep.155

April 11th

can't hold back laughter
웃참 실패
ut-cham shil-pae

When V bursts into laughter at something SUGA does while reciting his handwritten six-line poem, the subtitle "**웃참 실패** ut-cham shil-pae" appears. "**웃참** ut-cham" is a new word that combines "**웃음** u-seum" (smile/laugh) and "**참기** cham-kki" (holding back), and "**실패** shil-pae" means "failure." You can use "**웃참 실패** ut-cham shil-pae" when you fail to hold back laughter. For instance, if you burst out laughing while watching <Run BTS!> on a quiet bus, you can say "**웃참 실패** ut-cham shil-pae!"

Don't stare at him. His face is turning red.

<Run BTS!> Ep.128

turn red (face)
얼굴이 빨개지다
eol-gu-ri ppal-gae-ji-da

Jin has become a liar while playing the Liar Game! He blushes as he makes up lies to convince the others that he knows the keyword. When your face (얼굴 eol-gul) turns red from shyness or embarrassment, you can use the word "빨갛다 ppal-ga-ta" (to be red) to make the expression "얼굴이 빨개지다 eol-gu-ri ppal-gae-ji-da." Have you ever imagined making eye contact with BTS? 얼굴이 빨개져 eol-gu-ri ppal-gae-jeo! 얼굴이 빨개져요 eol-gu-ri ppal-gae-jeo-yo!

04

As expected, does fortune favor me?

<Run BTS!> Ep.95

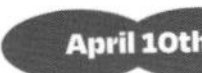

Fortune favors me.

하늘은 나의 편

ha-neu-reun na-ui pyeon

Jung Kook is playing hopscotch with Jimin. When Jimin makes a mistake, Jung Kook says, "하늘은 나의 편 ha-neu-reun na-ui pyeon," which literally means, "Heaven is on my side." When heaven (하늘 ha-neul) is on your side (편 pyeon), it feels like fortune favors you because everything is going well. When things go well, try using the expression "하늘은 나의 편 ha-neu-reun na-ui pyeon!"

[BTS VLOG] Jimin I 팔찌공방 VLOG

September 20th

have slow hands

손이 느리다

so-ni neu-ri-da

Jimin is at a bracelet workshop and is concerned that he might keep others waiting because his hands (손 son) are slow (느리다 neu-ri-da). "**손이 느리다** so-ni neu-ri-da" is an expression used when someone works slowly either because they are unfamiliar with the task or because they want to do things perfectly. On the contrary, if you are a swift worker, "**손이 빠르다** so-ni ppa-reu-da" is how you would describe yourself.

과연 예능의 신은 윤기의 편을 들 것인지

Will the god of entertainment really help SUGA?

<Run BTS!> Special Episode – Next Top Genius Part 2

April 9th

the god of entertainment
예능의 신
ye-neung-ui shin

It's SUGA's turn in a dice game, and he's currently in the lead! However, he decides that it would be more fun to lose his lead than to maintain it as expected. Before rolling the dice, he wonders whether "예능의 신 ye-neung-ui shin" will help him or not. "예능 ye-neung" refers to entertainment programs, and "신 shin" means "a god." Since entertainment programs aim to make viewers laugh, SUGA hopes that "예능의 신 ye-neung-ui shin" will assist him in amusing everyone with his unexpected strategy.

09

V, you're sharrrp.

<Run BTS!> Special Episode
– 'RUN BTS TV' On-air Part 1

September 21st

H: 눈썰미 좋아요.
nun-sseol-mi jo-a-yo

You're sharp.

눈썰미 좋다.

nun-sseol-mi jo-ta

V says he thinks he recognizes the drum instructor invited to <Run BTS!>. He suspects he may have seen him on another program. When it turns out that V is right, Jin compliments him by saying "**눈썰미 좋다** nun-sseol-mi jo-ta." This means that you have a good ability to notice or copy something just by looking at it once. Jin slightly raises his intonation, saying, "**눈썰미 좋다잉** nun-sseol-mi jo-ta-ing?" The addition of "**잉** ing" at the end of the sentence is one of the features of a dialect accent in Korean, but now it's often used for fun!

04

Astonished

<Run BTS!> Special Episode
- Telepathy Part 1

astonished

kkam-nol

When V is surprised to hear that he must wear a blindfold and move to an unknown place, the subtitle says "깜놀 kkam-nol." This expression is a shortened form of "깜짝 놀라다 kkam-jjak nol-la-da." "깜짝 kkam-jjak" describes the sudden movement of someone surprised, and "놀라다 nol-la-da" (☛ January 17th) means "to be startled." If BTS were to release new songs without prior notice, you could shout, "깜놀 kkam-nol!"

09

We all eat like a bird.

<Run BTS!> Ep.154

We eat like a bird.

입이 짧아요.

i-bi jjal-ba-yo

BTS ate 15 servings of food! Jimin thinks they ate less food than he expected, so he says "**입이 짧아요** i-bi jjal-ba-yo." "**입** ip" means "mouth" and "**짧다** jjal-tta" means "to be short." Therefore, "**입이 짧다** i-bi jjal-tta" is an idiomatic expression that refers to a picky eater or a light eater. As BTS is known for their huge appetites, it's clear that Jimin is making a joke, right?

07

<Run BTS!> Special Episode
– Telepathy Part 1

amazing
신기방기
shin-gi-bang-gi

BTS is making a toast, cheerfully shouting, “Run BTS!” When Jung Kook finds out that the drink is non-alcoholic, the subtitles say “**신기방기** shin-gi-bang-gi.” This expression is a new term that refers to an interesting and enjoyable state created by the discovery of something unbelievable. “**신기** shin-gi” comes from “**신기하다** shin-gi-ha-da” (👉 July 14th), the expression used when seeing something unusual and amazing. It is followed by “**방기** bang-gi” to make a rhyme. When you visit BTS’ newest pop-up store full of amazing merchandise, you can say “**신기방기해** shin-gi-bang-gi-hae!” or “**신기방기해요** shin-gi-bang-gi-hae-yo!”

09

23

Wow, I got goosebumps!!

<Run BTS!> Ep.154

September 23rd

H: 소름 돋았어요.
so-reum do-da-sseo-yo

I got goosebumps.
소름 돋았어.
so-reum do-da-sseo

RM declares that today's mission is literally running to their company in Yongsan. j-hope believes him and is very surprised. When he realizes that it was a joke, he exclaims, "**소름 돋았어** so-reum do-da-sseo." "**소름** so-reum" usually refers to goosebumps on the skin induced by surprise or the cold. If you watch BTS' mesmerizing choreography in person, you might exclaim the expression "**소름 돋았어** so-reum do-da-sseo!" or "**소름 돋았어요** so-reum do-da-sseo-yo!" to express your excitement and awe.

04

We're all for it.

<Run BTS!> Ep.148

We're all for it.

적극 찬성

jeok-geuk chan-seong

During the filming of <Run BTS!>, BTS is divided into two teams. RM suggests that the two oldest members, Jin and SUGA, should be the leaders. When the others agree, the subtitle "**적극 찬성** jeok-geuk chan-seong" appears on the screen. "**적극** jeok-geuk" refers to an active attitude toward something, and "**찬성** chan-seong" means "agreement." In other words, "**적극 찬성** jeok-geuk chan-seong" means that someone completely agrees with something. If you disagree, you can say "**반대** ban-dae." What if someone suggests listening to BTS songs all day today? **적극 찬성** jeok-geuk chan-seong!

My stomach hurts.

<Run BTS!> Ep.150

My stomach hurts.

아이코 배야.

a-i-ko bae-ya

Jimin and V are taking a quiz about proverbs! When Jimin gives a funny answer, the others burst into laughter. At this time, the subtitles say "아이코 배야 a-i-ko bae-ya." "아이코 a-i-ko" is a word that represents the sound made when someone is in pain, surprised, happy or having fun. It's similar to the exclamation "아이고 a-i-go" (☛ November 5th). Doesn't your stomach hurt when you laugh too much? That's why they say "아이코 배야 a-i-ko bae-ya!" in this context. If a particular scene on <Run BTS!> is so hilarious that you roll on the floor, use the expression "아이코 배야 a-i-ko bae-ya!"

<Run BTS!> Ep.141

Let's do pop songs this time!!

April 5th

Let's go.

가즈아.

ga-jeu-a

During a game where BTS tries to guess songs from a particular genre, j-hope suggests they choose pop music. The subtitle "가요로 가즈아 ga-yo-ro ga-jeu-a" appears, with "가요 ga-yo" meaning "pop song" and "가즈아 ga-jeu-a" derived from the phrase "가자 ga-ja" which means "Let's go." When "가자 ga-ja" is pronounced slowly, it sounds like "가즈아 ga-jeu-a." It is an expression that is playfully used in a comfortable situation with close friends to convey positive expectations or desires for something. For example, if you hope a new BTS song ranks first on the charts, you can say "방탄소년단 1등 가즈아 bang-tan-so-nyeon-dan il-tteung ga-jeu-a" using "1등 il-tteung" (first place 👉 June 24th).

25

Jung Kook used his head well in a tricky situation.

<Run BTS!> Ep.135

use one's head

머리를 쓰다

meo-ri-reul sseu-da

BTS is playing a game where they must hide something. Jung Kook hides the object in the room that they have just been in. At that moment, "머리를 쓰다 meo-ri-reul sseu-da" appears on the screen, suggesting that he is using his head. This expression means "to think rationally about something or to come up with a brilliant idea." It is commonly used with "잘 jal," which means "well." So, you can say "머리를 잘 쓰다 meo-ri-reul jal sseu-da."

04

childlike palate
아기 입맛
a-gi im-mat

While filming <Run BTS!>, RM calls Jin "**아기 입맛** a-gi im-mat" because Jin likes drinks with chunks of gelatin in them. This expression is a combination of "**아기** a-gi" (baby) and "**입맛** im-mat" (taste in food), so it refers to the palate of babies who prefer sweet flavors to strong ones. Having a palate suited for strong-tasting food is called "**어른 입맛** eo-reun im-mat," using the word "**어른** eo-reun," which means "adult."

Two gullible men who are unhappy

<Run BTS!> Ep.144

September 26th

gullible

귀가 얇다

gwi-ga yal-tta

BTS tries to guess their number one summer song chosen by ARMY! At first, j-hope and V correctly guess the answer, but they are persuaded by Jimin to change it at the last minute, and unfortunately, they get it wrong. In the subtitles, they are described as "**귀가 얇다** gwi-ga yal-tta." This idiomatic expression combines "**귀** gwi" (ear) and "**얇다** yal-tta" (to be thin). It refers to someone who is easily influenced or deceived by another. Do you find yourself easily swayed like j-hope and V?

04

Getting far ahead of themselves

<Run BTS!> Ep.136

drinking kimchi soup (getting ahead of oneself)

김칫국 드링킹

gim-chit-guk deu-ring-king

Before starting a nonsense quiz, Jin and j-hope feel confident that they have already won. The subtitles say "**김칫국 드링킹** gim-chit-guk deu-ring-king," which comes from a Korean proverb "**김칫국부터 마신다** gim-chit-guk-bu-teo ma-shin-da," which means "Don't count your chickens before they hatch." "**김칫국** gim-chit-guk" means "kimchi soup," and "**드링킹** deu-ring-king" is the Korean spelling of the English word "drinking," which means "**마시다** ma-shi-da." Therefore, "**김칫국 드링킹** gim-chit-guk deu-ring-king" is used between close friends for fun when a person assumes something they want to happen will happen.

09

27

[BTS VLOG] V I DRIVE VLOG

September 27th

H: 뜨끈뜨끈해요.
tteu-kken-tteu-kkeun-hae-yo

It's nice and toasty.

뜨끈뜨끈해.

tteu-kkeun-tteu-kkeun-hae

V buys some freshly cooked corn and returns to his car, exclaiming "뜨끈뜨끈해 tteu-kkeun-tteu-kkeun-hae" in admiration. This expression refers to something that is steaming hot. It can be used when the temperature of something is quite high, such as when you have a fever or when a room is warm due to good heating. When you get into a bathtub filled with hot water, you can use this expression "뜨끈뜨끈해 tteu-kkeun-tteu-kkeun-hae" or "뜨끈뜨끈해요 tteu-kkeun-tteu-kkeun-hae-yo" to describe the hot temperature of the water.

<Run BTS!> Ep.142

April 2nd

No matter what, it's iced Americano.

얼죽아

eol-ju-ga

RM calls himself a member of the "얼죽아 eol-ju-ga" association! The term is an abbreviation of "얼어 죽어도 아이스 아메리카노 eo-reo ju-geo-do a-i-seu a-me-ri-ka-no," which refers to people who enjoy drinking iced Americanos even in freezing weather. In Korea, there are many such people, so RM has jokingly created a fictional group called the "얼죽아 eol-ju-ga" association and pretends to be a member.

09

Looking around

<Run BTS!> Ep.148

looking around

두리번두리번

du-ri-beon-du-ri-beon

Jin is looking around to find the right pair of gloves that fit before starting the interior construction! At this time, the subtitles read: **두리번두리번** du-ri-beon-du-ri-beon. This expression describes someone looking around. Do you look around to find your seat when you go to a BTS concert? You can describe what you're doing with "**두리번두리번** du-ri-beon-du-ri-beon."

04

mukbang (eating broadcast)

먹방

meok-bang

While Jin takes on the role of a teacher and conducts a lecture, RM and j-hope sneakily eat instant cup ramyeon behind notebooks standing on their desk. The subtitle describes the situation as "먹방 meok-bang." This is a new term that combines "먹 meok," which means "eating" (먹다 meok-da), and "방 bang," which comes from the word "broadcast" (방송 bang-song). Therefore, it can be defined as "eating broadcast." Do you remember when Jin created a *mukbang* series called "EAT Jin?" When you feel hungry and it's too late to eat, satisfy your hunger by watching a BTS "먹방 meok-bang."

Binge-eating in *mukbang*

<Run BTS!> Ep.113

09

Whispering

<Run BTS!> Ep.128

whispering
속닥속닥
sok-dak-sok-dak

V whispers to j-hope, and the subtitle describes this as "속닥속닥 sok-dak-sok-dak." This expression is used to describe someone whispering in a low voice. You can use "속닥 sok-dak" multiple times to describe someone whispering for a long time. Let me tell you a secret. 속닥속닥속닥 sok-dak-sok-dak-sok-dak...

4월

sa-wol

April

09

30

j-hope
굉장히 명절만 오면 설렙니다

I'm thrilled when the holidays come.

[EPISODE] BTS (방탄소년단)
2021 'DALMAJUNG' Shoot

September 30th

I'm thrilled.

설렙니다.

seol-lem-ni-da

Just before *Chuseok*, which is Korean Thanksgiving, j-hope is thrilled and looking forward to eating all the holiday treats. So he says "**설렙니다** seol-lem-ni-da," which is a conjugation of "**설레다** seol-le-da," an expression used when you feel thrilled and your heart flutters. You can casually say "**설레** seol-le" to your close friends. If you feel elated at a BTS concert tomorrow, try saying "**설레** seol-le" or "**설렙니다** seol-lem-ni-da" to express how excited you are!

What happened? lol

<Run BTS!> Special Episode
- Mini Field Day Part 1

What happened?

이게 무슨 일?

i-ge mu-seun il

Due to SUGA mistiming Jung Kook's serve, their volleyball game ends in disappointment. The subtitles say, "이게 무슨 일 i-ge mu-seun il," which means "What happened?" This phrase can be used as an exclamation when you don't understand how something happened, you can't believe something, or you're surprised. If a BTS live video stream suddenly cuts off, leaving you with a blank screen, you can say "이게 무슨 일 i-ge mu-seun il?"

10월

shi-wol

October

03

30

감이 좋아.
ga-mi jo-a
HELLO 2021
슈가의 선택은 지민~!

SUGA chooses Jimin!

* SUGA did not attend due to personal reasons.
<Run BTS!> Ep.128

H: 감이 좋아요.
ga-mi jo-a-yo

He has a good intuition.

감이 좋아.

ga-mi jo-a

While playing the Liar Game, RM and Jimin are suspected of being liars! V calls SUGA, who is not in the studio. Without explaining the situation, V asks him to choose between RM and Jimin. When SUGA picks Jimin, RM describes him by saying, "감이 좋아 ga-mi jo-a." As "감 gam" refers to "a feeling or sense," "감이 좋다 ga-mi jo-ta" means that someone is good at predicting or noticing something. If you have a friend with a great intuition, you can use the expression: 감이 좋아 ga-mi jo-a! 감이 좋아요 ga-mi jo-a-yo!

Rewinding seven seconds. This is sweet.

<Run BTS!> Special Episode
- 'RUN BTS TV' On-air Part 2

October 1st

This is sweet.

꿀입니다.

kku-rim-ni-da

During his own live video stream, RM talks about the ability to reverse time. He uses the expression "꿀입니다 kku-rim-ni-da" and adds that with this ability, you can instantly undo your mistakes. When something is convenient and easy, it is referred to as "꿀 kkul" (honey), which is sweet, tasty, and good for your health. You can casually say "꿀이야 kku-ri-ya" to your close friends. If you have discovered an easy way to do something, try saying "꿀이야 kku-ri-ya" or "꿀입니다 kku-rim-ni-da" to express how easy and convenient it is.

03

Compelling

<Run BTS!> Ep.150

compelling

흥미진진

heung-mi-jin-jin

BTS must guess the meaning of some new abbreviations! When SUGA tries to infer the meaning of a word, Jin and RM watch him with interest. The subtitle "흥미진진 heung-mi-jin-jin" appears on the screen. This expression refers to a situation in which something interesting draws your attention. When BTS is playing a game in <Run BTS!>, and you can't take your eyes off it, try the expression "흥미진진해 heung-mi-jin-jin-hae" or "흥미진진해요 heung-mi-jin-jin-hae-yo."

10

02

꽃 선물 받고 심쿵

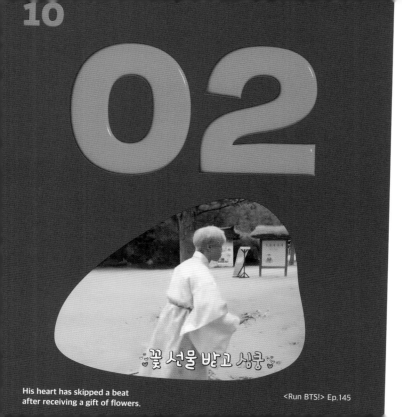

His heart has skipped a beat after receiving a gift of flowers.

<Run BTS!> Ep.145

October 2nd

heart skipping a beat

심쿵

shim-kung

During a treasure hunt, Jimin receives a gift of flowers from RM. RM nonchalantly says he picked them on his way, but Jimin is very moved. At that moment, Jimin's heart flutters with excitement and gratitude, which is expressed in the subtitles as "심쿵 shim-kung." This is a new expression that combines "심 shim," which means "heart" (심장 shim-jang), and "쿵 kung," which is the sound made when a large and heavy object falls onto the ground. In other words, your heart is fluttering so hard that it falls over with a "thump." When you watch fabulous live BTS videos, you can say "심쿵 shim-kung."

03

28

감 잡았어.
gam ja-ba-sseo
HELLO 2021
이제 감을 잡아버렸다!

He has now got the hang of it!

* SUGA did not attend due to personal reasons.
<Run BTS!> Ep.128

H: 감 잡았어요.
gam ja-ba-sseo-yo

I got the hang of it.
감 잡았어.
gam ja-ba-sseo

BTS is playing the Liar Game! By the third round, j-hope finally understands how to play the game and says, "감 잡았어 gam ja-ba-sseo." "감 gam" refers to "a feeling or sense," and "잡다 jap-da" means "to catch." So it means "to get a sense of or to learn how to do something." Are you becoming familiar with Korean? Then you can say "감 잡았어 gam ja-ba-sseo!" or "감 잡았어요 gam ja-ba-sseo-yo!"

10

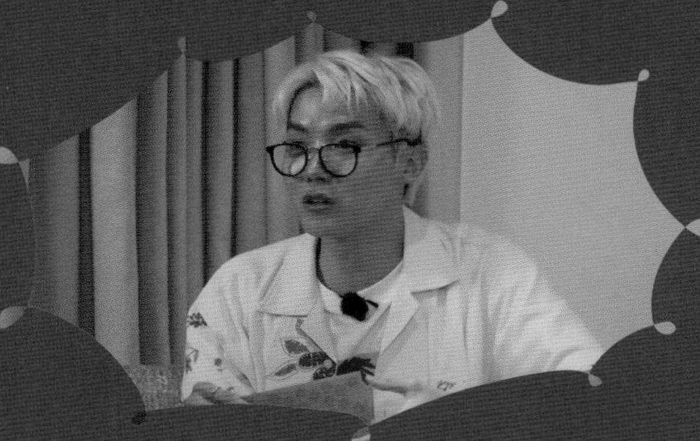

<Run BTS!> Ep.151

October 3rd

H: 대박이에요.
dae-ba-gi-e-yo

It's a jackpot.
대박이다.
dae-ba-gi-da

The staycation in an episode of <Run BTS!> is over, and it's time to receive the reward. j-hope says, "**대박이다** dae-ba-gi-da" because he's surprised that the stakes are very high. This expression is used when earning a huge, unexpected profit, or when you're in awe. You can simply say "**대박** dae-bak" to express your surprise. When BTS performance is unbelievable, try saying "**대박** dae-bak," "**대박이다** dae-ba-gi-da," or "**대박이에요** dae-ba-gi-e-yo."

Confident

<Run BTS!> Special Episode
- Telepathy Part 1

March 27th

confident

자신만만

ja-shin-man-man

BTS is given a description of a situation, and they all must strike the same pose to illustrate it! After listening to the game rules, SUGA says they are good at this type of game. The subtitle describes him as "자신만만 ja-shin-man-man." This expression refers to a state of overflowing confidence. When with BTS, ARMY is always 자신만만 ja-shin-man-man!

10

04

The kindest person in the world...

<Run BTS!> Ep.154

October 4th

the kindest person in the world
세상 친절
se-sang chin-jeol

While pouring a drink for RM and answering his question about the drink, V is captioned with "**세상 친절** se-sang chin-jeol," which means that he is kinder than anyone else in the world. If a noun comes after "**세상** se-sang" (world), it creates a new expression referring to someone being the best in the world at something. For example, if you think BTS is cool, you can use "**멋짐** meot-jim" to say "**세상 멋짐** se-sang meot-jim," right?

03

Checking with nervous excitement...

<Run BTS!> Ep.135

nervous yet excited

걱정 반 기대 반

geok-jeong ban gi-dae ban

Jin has to hide a randomly assigned object, and he grows nervously excited as he waits to find out what the item is. The subtitles describe him as "걱정 반 기대 반 geok-jeong ban gi-dae ban," which literally means that someone is half concerned and half excited. This expression can be used when someone feels anxious and giddy at the same time. When you travel to Korea for the first time, you can express your emotions with "걱정 반 기대 반 geok-jeong ban gi-dae ban."

10

Don't you know the saying, "crispy outside, soft inside?" <Run BTS!> Ep.131

October 5th

crispy outside, soft inside

겉바속촉

geot-ba-sok-chok

RM uses the expression "**겉바속촉** geot-ba-sok-chok" to describe a peach, which is hard on the outside and soft on the inside. The expression is originally used to describe crispy food that is tender on the inside. It is an acronym of "**겉은 바삭하고 속은 촉촉하다** geo-teun ba-sa-ka-go so-geun chok-cho-ka-da" (**바삭** ba-sak 👉 July 23rd). The phrase is often associated with fried food such as fried chicken or *tangsuyuk* (sweet and sour pork), but you can use it to describe a person who looks cold-hearted but actually has a warm heart.

03

Jung Kook can handle it on his own.

<Run BTS!> Ep.149

March 25th

I can handle it on my own.

혼자서도 잘해요.

hon-ja-seo-do jal-hae-yo

SUGA is about to assist Jung Kook in assembling a bookshelf. However, Jung Kook insists on doing it by himself as he believes he can handle it alone. The subtitles describe this situation, using "혼자서도 잘해요 hon-ja-seo-do jal-hae-yo." This expression is used when you are capable of doing something on your own, like Jung Kook. Is there something that you can do well by yourself? Then give yourself a compliment using "난 nan" or "전 jeon," which both mean "I." 난 혼자서도 잘해 nan hon-ja-seo-do jal-hae! 전 혼자서도 잘해요 jeon hon-ja-seo-do jal-hae-yo!

10

06

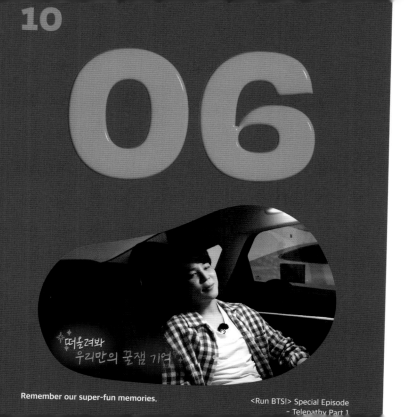

떠올려봐
우리만의 꿀잼 기억

Remember our super-fun memories.

<Run BTS!> Special Episode
- Telepathy Part 1

super fun

꿀잼

kkul-jaem

BTS is on a mission to find their own "꿀잼 장소 kkul-jaem jang-so," which means "super-fun places." "꿀잼 kkul-jaem" is a new word that combines "꿀 kkul" (honey 👉 October 1st) and "재미 jae-mi" (fun). Honey implies something that is sweet and pleasurable, and "재미 jae-mi" is shortened to one syllable: 잼 jaem. This expression can be used to describe something extremely fun. If an episode of <Run BTS!> cracks you up, you can say "꿀잼 kkul-jaem!"

03

March 24th

We can't see ahead.

한 치 앞도 모른다.

han chi ap-do mo-reun-da

In an interview, j-hope talks about his future and says "한 치 앞도 모른다 han chi ap-do mo-reun-da." In this expression, "치 chi" refers to a very short distance of 3.03 cm (1.2 in), but it is not commonly used these days. So, it means "We can't even see 3.03 cm ahead," meaning that in life, we can't anticipate what will happen. Who could have imagined that you would be studying Korean like this with BTS? In this sense, life is 한 치 앞도 모른다 han chi ap-do mo-reun-da, isn't it?

10

07

blood, sweat, and tears

피땀눈물 (p;ㅠ)

pi-ttam-nun-mul

The BTS song <Blood Sweat & Tears> can be represented as "p;ㅠ" for fun. The English letter "p" has the same pronunciation as the Korean word "피 pi" (blood). A semicolon is commonly used as a symbol of "땀 ttam" (sweat) in Korean online chat rooms, as it looks like dripping sweat. Finally, the Korean vowel "ㅠ" looks like someone dropping "눈물 nun-mul" (tear) as in "ㅠㅠ" or "ㅜㅜ." When combined, it becomes "blood, sweat, and tears" or "p;ㅠ!"

03

Appealing his status as a beginner

<Run BTS!> Ep.129

March 23rd

This is the first time in my life.

난생처음입니다.

nan-saeng-cheo-eu-mim-ni-da

SUGA is learning tennis for the first time in his life. After watching Jin display a decent performance, SUGA emphasizes that he is a novice and says "난생처음입니다 nan-saeng-cheo-eu-mim-ni-da." This expression is used to indicate that it is the first time you have done something since you were born. Try saying "난생처음이야 nan-saeng-cheo-eu-mi-ya" or "난생처음입니다 nan-saeng-cheo-eu-mim-ni-da" when you are doing something for the first time in your life.

08

This is exactly my style.

<Run BTS!> Ep.123

H: 취향저격이에요.
chwi-hyang-jeo-gyeo-gi-e-yo

This is exactly my style.
취향저격이다.
chwi-hyang-jeo-gyeo-gi-da

Jung Kook tries a kimchi quesadilla for the first time and says "**취향저격이다** chwi-hyang-jeo-gyeo-gi-da." When something is your style (**취향** chwi-hyang), you can use this expression. You can also use it when you meet someone who is your ideal type of person, instead of a song, movie, or food. If you hear a song on the street and fall in love with it, try using this expression: **취향저격이다** chwi-hyang-jeo-gyeo-gi-da! **취향저격이에요** chwi-hyang-jeo-gyeo-gi-e-yo!

03

Huhu... You're kidding.

<Run BTS!> Ep.130

March 22nd

a joke

농담

nong-dam

Before a tennis match with V, j-hope says that V will win! V takes it as a joke with a smile, with "농담 nong-dam" (joke) appearing next to him in the subtitles. Since j-hope is likely to win, V treats it as a jest. After you make a joke to someone, try saying "농담이야 nong-da-mi-ya" or "농담이에요 nong-da-mi-e-yo."

A face is a weapon.

얼굴이 무기

eol-gu-ri mu-gi

Jin uses a witty expression, "얼굴이 무기 eol-gu-ri mu-gi," during a live video stream for his birthday. It literally means that a face (얼굴 eol-gul) is a weapon (무기 mu-gi). In this case, "weapon" is a metaphor for strengths that others find charming. You can use "N (noun)이/가 무기 i/ga mu-gi" to highlight your strong points. If you have an attractive voice (목소리 mok-so-ri), you can say "목소리가 무기 mok-so-ri-ga mu-gi."

* Nouns ending with a batchim use "이 i," and nouns ending without one use "가 ga."

03

21

<Run BTS!> Ep.106

March 21st

appearing and disappearing

보여줄랑 말랑

bo-yeo-jul-lang mal-lang

During a photo exhibition in an episode of <Run BTS!>, Jimin declares that his partially revealed face is the highlight of a photo. If someone is hesitant to show you something, you can use the expression "보여줄락 말락 bo-yeo-jul-lak mal-lak," which is derived from the verb "보여주다 bo-yeo-ju-da" (to show). Jimin changed the batchim "ㄱ" in "락 lak" to "ㅇ," and said "보여줄랑 말랑 bo-yeo-jul-lang mal-lang" to make it sound cuter.

10

10

만약에...
여기서 누구 한 명 지더라도

젔잘싸
jeot-jal-ssa

8 9

TENNIS
RUN BTS

Even if either of them loses here

<Run BTS!> Ep.130

lost, but well fought

졌잘싸

jeot-jal-ssa

RM and V are playing tennis! They're neck-and-neck (**막상막하** mak-sang-ma-ka 👉 June 28th), and the game is at match point. j-hope says "**졌잘싸** jeot-jal-ssa," which is an acronym of the expression "**졌지만 잘 싸웠다** jeot-ji-man jal ssa-wot-da," which means "It was a well-fought match even if whoever lost." Try saying "**졌잘싸** jeot-jal-ssa" when someone loses a tight match on <Run BTS!>.

03

20

과연 그들은 척하면 척할 수 있을지?!

Can they be in sync?!

<Run BTS!> Ep.97

March 20th

in sync

척하면 척

cheo-ka-myeon cheok

BTS is about to play a telepathy game! j-hope emphasizes the rock-solid teamwork they have had for a long time, saying, "척하면 척 cheo-ka-myeon cheok." This expression is used when people close to one another can seemingly read each other's minds and are always in sync. Because BTS and ARMY have spent plenty of time together, we know what's on each other's minds, right? 척하면 척 cheo-ka-myeon cheok!

Mussels are the best!

<Run BTS!> Ep.154

H: 짱이에요.
jjang-i-e-yo

It's the best.

짱이다.

jjang-i-da

While tasting different kinds of food, Jung Kook says, "홍합이 짱이다 hong-ha-bi jjang-i-da," which means "Mussels are the best." "짱 jjang" is an idiomatic expression used to refer to something that is "the best," "top," or "number one." When SUGA gives a thumbs up and says "캡짱 kaep-jjang," the word "짱 jjang" has the same meaning. "캡 kaep," which comes from the word "captain" in English, is slang meaning "the best." Try saying "짱이다 jjang-i-da!" or "짱이에요 jjang-i-e-yo!" when talking about BTS' new release.

Does the goddess of luck really favor j-hope?

<Run BTS!> Ep.130

I just got lucky.

그저 운일 뿐이에요.

geu-jeo u-nil ppu-ni-e-yo

During a tennis match, j-hope suddenly starts playing better! When the others ask if he hadn't been trying his best before, he says, "그저 운일 뿐이에요 geu-jeo u-nil ppu-ni-e-yo." This expression means "I just got lucky," and is used when something good happens by chance or works out without any special effort. For example, if you receive a good grade even though you didn't study hard, you can say "그저 운일 뿐이야 geu-jeo u-nil ppu-ni-ya," or "그저 운일 뿐이에요 geu-jeo u-nil ppu-ni-e-yo."

12

빵 터짐

Bursting into laughter

<Run BTS!> Ep.149

bursting into laughter

ppang teo-jim

In an episode of <Run BTS!> about interior design, Jin bursts into laughter after finding that someone wiped off the paint on the tile he was supposed to use. "빵 터짐 ppang teo-jim" is an expression used when something cracks you up. "빵 ppang" is the sound made when something like a balloon or bomb explodes, and "터지다 teo-ji-da" means "to explode." They form a new expression that means "bursting into laughter." When <Run BTS!> makes you laugh out loud, try saying "빵 터짐 ppang teo-jim!"

18

The master of cramming

<Run BTS!> Ep.146

cramming
벼락치기
byeo-rak-chi-gi

V has been tasked with memorizing and reciting a *sijo*, a traditional Korean poem from the *Joseon* period. As he practices diligently, the subtitle "벼락치기 byeo-rak-chi-gi" appears on the screen. This expression comes from the word "벼락 byeo-rak," which means "thunderbolt," and is used to describe doing something in a hurry at the last minute, like a sudden bolt of thunder. If you find yourself needing to finish an important task after a binge-watching session of <Run BTS!>, you can use the expression "벼락치기 byeo-rak-chi-gi" to describe your last-minute efforts.

October 13th

13

Happy Birthday, Jimin!

생일 축하해요, 지민!

saeng-il chu-ka-hae-yo, Jimin

03

Confident

<Run BTS!> Ep.133

eating cold rice porridge (a piece of cake)

식은 죽 먹기

shi-geun juk meok-gi

Before starting a dancing game, j-hope confidently demonstrates that he understands the rules by saying "식은 죽 먹기 shi-geun juk meok-gi." "죽juk" (rice porridge) is a food that is soft and easy to digest, even if not chewed it much. When "죽juk" is cooked and allowed to cool to a suitable temperature before eating, it becomes even easier and more comfortable to eat. That's why we use the expression "식은 죽 먹기shi-geun juk meok-gi" to describe something that is very easy to do. Learning Korean with BTS? 식은 죽 먹기 shi-geun juk meok-gi!

Jin hyung, perfect!

<Run BTS!> Ep.125

October 14th

Perfect.

완벽해요.

wan-byeo-kae-yo

Jin is cooking *ramyeon* (Korean instant noodles) with ham. His dish seems to both look and taste perfect, and j-hope compliments him by saying "완벽해요 wan-byeo-kae-yo!" You can casually say "완벽해 wan-byeo-kae" to your close friends. When BTS has a perfect performance on stage, try to use this expression: 완벽해 wan-byeo-kae! 완벽해요 wan-byeo-kae-yo!

03

16

<Weverse Live> 2021.02.19

March 16th

a wisdom tooth
사랑니
sa-rang-ni

During a live video stream, when asked about his New Year's Day activities, SUGA says that he had his "**사랑니** sa-rang-ni" (wisdom tooth) extracted. Wisdom tooth is the last molar tooth that usually appears between a person's late teenage years and early 20s. It is called "**사랑니** sa-rang-ni" because some believe it grows at the age when one can understand love, which means "**사랑** sa-rang," while others say that it can be very painful, much like your first heartbreak.

10

15

잘한다!

Good job!

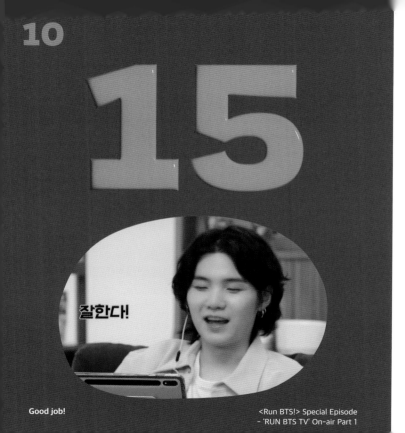

<Run BTS!> Special Episode
– 'RUN BTS TV' On-air Part 1

H: 잘해요.
jal-hae-yo

Good job.

잘한다.

jal-han-da

While looking at j-hope, who has become a temporary kids content creator, SUGA compliments him by saying "**잘한다** jal-han-da!" This expression is used as praise when someone accomplishes a task perfectly. When BTS is singing, dancing, or playing a game, try saying this expression: **잘한다** jal-han-da! **잘해요** jal-hae-yo!

03

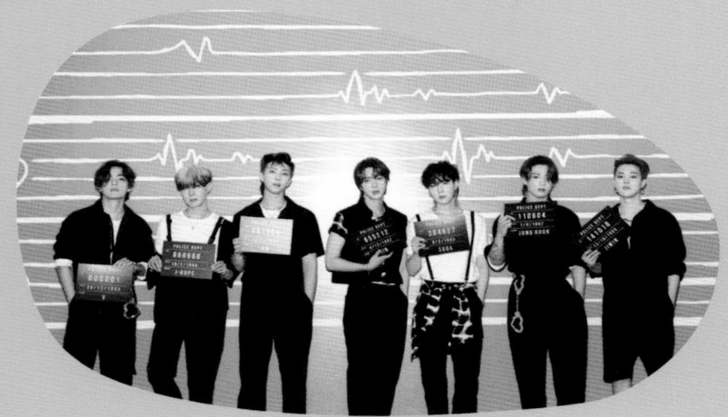

The day has finally come.

드디어 그날이 왔어요.

deu-di-eo geu-na-ri wa-sseo-yo

BTS is back with <Butter>! They are delighted to finally reveal the song to ARMY. j-hope says, "**드디어 그날이 왔어요** deu-di-eo geu-na-ri wa-sseo-yo," which implies that the release date has come at last. In this sentence, "**드디어** deu-di-eo" is an expression indicating that something you have longed for has finally come true. On the day of a BTS concert you have been waiting for, you can say "**드디어 그날이 왔어** deu-di-eo geu-na-ri wa-sseo!" or "**드디어 그날이 왔어요** deu-di-eo geu-na-ri wa-sseo-yo!"

10

16

BTS (방탄소년단) MBTI Lab 1

October 16th

H: 대단해요.
dae-dan-hae-yo

Incredible.
대단해.
dae-dan-hae

While SUGA prefers to stay home on his day off, RM likes to wander to different places. SUGA admires RM's enthusiasm and says, "**대단해** dae-dan-hae," which means "Incredible." Doesn't BTS seem incredible when they perform difficult choreography, while singing on stage? Try saying "**대단해** dae-dan-hae!" or "**대단해요** dae-dan-hae-yo!" when you watch their performance.

03

Jimin
7년 동안 옆에서 응원해주는 일은 쉬운 일이 아닙니다

It must not have been an easy thing to keep supporting us for seven years.

[2020 FESTA] BTS (방탄소년단) '방탄생파'

March 14th

It's not an easy thing.

쉬운 일이 아닙니다.

shwi-un i-ri a-nim-ni-da

Jimin says ARMY has been supporting BTS for seven years, adding "**쉬운 일이 아닙니다** shwi-un i-ri a-nim-ni-da," which means "It's not an easy thing." Making history by breaking records is no "**쉬운 일** shwi-un il" (easy thing) for BTS either. Nevertheless, BTS and ARMY continue doing these difficult things because they exist for each other!

10

Cutie

<Run BTS!> Ep.142

October 17th

cutie

애교쟁이

ae-gyo-jaeng-i

Jimin is so excited to try *jjajang ramyeon* (instant noodles in black bean sauce) that he begins to hum and dance! The subtitle uses "애교쟁이 ae-gyo-jaeng-i" to describe him. This expression is made up of "애교 ae-gyo," which means "the attempt to look cute to others through voice, facial expression, and tone," and "쟁이 jaeng-i," which refers to a person who has a lot of a specific quality. Therefore, it means "a person who acts cute." Likewise, you can say "멋쟁이 meot-jaeng-i" to someone who has a lot of "멋 meot" (style). 방탄소년단 멋쟁이 bang-tan-so-nyeon-dan meot-jaeng-i!

03

13

(고민보다 Go)
고민하면 안 돼.
go-min-ha-myeon an dwae

H: 고민하면 안 돼요.
go-min-ha-myeon an dwae-yo

Don't think too much.

고민하면 안 돼.

go-min-ha-myeon an dwae

BTS is taking a personality test! SUGA advises the others not to overthink during this type of test, saying, "고민하면 안 돼 go-min-ha-myeon an dwae." He says this because answering without hesitation leads to more accurate results when taking personality tests. This is similar to the BTS song <Go Go>. If someone is struggling to make a decision or is agonizing over something for a long time, try saying "고민하면 안 돼 go-min-ha-myeon an dwae," or "고민하면 안 돼요 go-min-ha-myeon an dwae-yo."

10

18

Let's take another look at the best scene of all time! <Run BTS!> Ep.94

October 18th

the best of all time

역대급

yeok-dae-kkeup

While BTS is playing the 007 Game, Jimin gets embarrassed for shouting late during his turn. Everybody laughs at that hilarious moment. As this scene is considered to be one of the funniest of <Run BTS!>, there are subtitles that read "역대급 명장면 yeok-dae-kkeup myeong-jang-myeon," which means "the best scene of all time." "역대급 yeok-dae-kkeup" is a new word for describing something that is the best or worst in history. If you think BTS' new release is the best song ever, try saying "역대급 yeok-dae-kkeup!"

03

12

그냥 뭐 재밌는 시간 보냈습니다 게임하면서

I just had a fun time playing games.

<Run BTS!> Ep.153

I had a fun time.

재밌는 시간 보냈습니다.

jae-min-neun shi-gan bo-naet-seum-ni-da

Jin is having so much fun recording <Run BTS!>. Near the end, he gives his thoughts, saying, "**재밌는 시간 보냈습니다** jae-min-neun shi-gan bo-naet-seum-ni-da." "**시간(을) 보내다** shi-gan(eul) bo-nae-da" means "to spend time doing something," and "**재밌는** jae-min-neun" (fun) is added to express that the time was enjoyable. Try saying this expression after having a great time at a BTS concert! **재밌는 시간 보냈어** jae-min-neun shi-gan bo-nae-sseo! **재밌는 시간 보냈습니다** jae-min-neun shi-gan bo-naet-seum-ni-da!

10

19

October 19th

a sugar rush

당 충전

dang chung-jeon

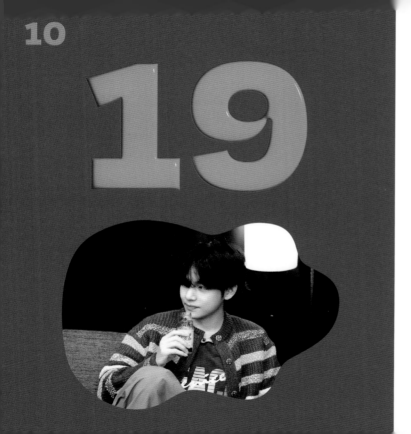

While preparing for a photo exhibition in an episode of <Run BTS!>, V drinks a sweet strawberry-banana drink to recharge his energy. "당 충전 dang chung-jeon" means "recharging your energy by consuming something sweet." When you're stressed out or tired, do you crave something sweet as V does? In that case, try "당 충전 dang chung-jeon" with your favorite sweets!

03

11

H: 즐겁게 삽시다.
jeul-geop-ge sap-shi-da

Let's live joyfully.

즐겁게 살자.

jeul-geop-ge sal-ja

In "77 Questions & 77 Answers," Jimin is asked about his motto, and he says, "즐겁게 살자 jeul-geop-ge sal-ja," which literally means "Let's live joyfully." If your motto is "Let's live happily," you can say "행복하게 살자 haeng-bo-ka-ge sal-ja." Do you have a purpose that drives your life? If you want to relish life as Jimin does, try this expression: 즐겁게 살자 jeul-geop-ge sal-ja!

10

20

Iced Americanos vs. hot Americanos

<Run BTS!> Ep.142

October 20th

iced Americanos vs. hot Americanos

아아 vs. 따아

a-a vs. tta-a

Koreans love Americanos! Everyone has a clear preference between "아이스 아메리카노 a-i-seu a-me-ri-ka-no" (iced Americanos) and "따뜻한 아메리카노 tta-tteu-tan a-me-ri-ka-no" (hot Americanos). They sometimes use "아아 a-a" or "따아 tta-a," the new words created from their initial letters. BTS unanimously prefers "아아 a-a!" Which do you prefer, "아아 a-a" or "따아 tta-a?"

<Run BTS!> Special Episode
- Telepathy Part 1

March 10th

one body, one soul

일심동체

il-sshim-dong-che

BTS is finding out how well they know each other in a game in an episode of <Run BTS!>. Before it starts, everyone takes turns saying something motivational, and V says, "일심동체 il-sshim-dong-che." This expression literally means "one body, one soul," and refers to a close relationship where mind and body are one. It can be used with old friends, lovers, family, and of course, BTS and ARMY!

10

21

Prankster

<Run BTS!> Ep.135

a prankster
장꾸
jang-kku

j-hope must guess the owner of a particular item! When the owner is about to be revealed, Jin stands up, acting like he is the man. The look on his face during the prank is just like "장난꾸러기 jang-nan-kku-reo-gi," which means a "a mischievous child." This expression is a combination of "장난 jang-nan" (prank) and "꾸러기 kku-reo-gi" (playful child). Nowadays, it is abbreviated to "장꾸 jang-kku." Try saying "장꾸 jang-kku" when describing your prankster friend!

March 9th

09

Happy Birthday, SUGA!

생일 축하해요, 슈가!

saeng-il chu-ka-hae-yo, SUGA

10

22

빼놓을 수 없는 치킨 + 맥주

Can't leave out chicken + beer

<Run BTS!> Ep.142

chicken and beer combo

chi-maek

BTS selects "치킨 chi-kin" (chicken) and "맥주 maek-ju" (beer) as one of the most fantastic pairings of Korean food! Koreans enjoy chicken and beer together most of the time and call it "치맥 chi-maek," a new abbreviation of the two words. This combination is commonly selected when having a picnic at Hangang Park or watching sports at home. Do you remember the <Run BTS!> episode where BTS eats "치맥 chi-maek" at a company gathering?

03

08

Should I give it a try?

<Run BTS!> Ep.126

H: 한번 해볼까요?
han-beon hae-bol-kka-yo

Should I give it a try?
한번 해볼까?
han-beon hae-bol-kka

Jung Kook keeps failing at a game where he must remove a cap without knocking down the bottle. Jimin suggests he try the game, saying, "한번 해볼까 han-beon hae-bol-kka?" In this expression, "한번 han-beon" means "as a trial," and "해보다 hae-bo-da" also means "to try," which further emphasizes the meaning of "to attempt." If you are hesitant to do something, you can say "한번 해볼까 han-beon hae-bol-kka?" to motivate yourself. You've got this!

10

23

BTS (방탄소년단) 'BE' Comeback Countdown

October 23rd

a magic touch

geum-son

BTS expresses their gratitude to ARMY for sending them heartfelt postcards. V admires the well-decorated postcards and remarks that many ARMY has "금손 geum-son," a new term derived from "금 geum" (gold) and "손 son" (hand), which means "skilled or talented with their hands," such as in cooking or drawing. Do you also have "금손 geum-son?"

03

<Run BTS!> Ep.151

March 7th

What are you doing?

뭐 해요?

mwo hae-yo

BTS is enjoying their vacation at a hotel! j-hope enters SUGA's room and asks him what he is doing. You can use the expression "뭐 해요 mwo hae-yo?" when asking someone what they are doing. Although it is in the honorific form, it can be used casually only in close relationships. "뭐 하세요 mwo ha-se-yo?" sounds more formal and polite. Meanwhile, you can simply say "뭐 해 mwo hae?" to your close friends.

Oh my God.

<Run BTS!> Ep.144

Oh my God!

heol

BTS is attempting to guess the title of a song based on its initial consonants in Korean. When Jung Kook easily guesses the correct answer, j-hope shows surprise. The subtitle says "헐 heol" to describe j-hope's astonishment. This is a new expression used when someone is astonished or shocked. If you ever see BTS in person at a concert and are amazed by their good looks, you can say "헐 heol" to express your surprise.

03

Everyone! We're running out of time!

<Run BTS!> Ep.120

March 6th

We're running out of time.

시간이 없어요.

shi-ga-ni eop-seo-yo

BTS is doing a role-play pretending to be in the 1970s! Jimin, who is immersed in his character, says "시간이 없어요 shi-ga-ni eop-seo-yo," telling the others to hurry up. This expression consists of "시간 shi-gan" (time) and "없다 eop-da" (do not exist). It means that there is no time to do something. It can be used to push someone, which is how Jimin uses it. When you need to go home and watch <Run BTS!>, but someone is moving slowly, try saying "시간이 없어 shi-ga-ni eop-seo" or "시간이 없어요 shi-ga-ni eop-seo-yo."

10

25

The older brothers are bewildered.

<Run BTS!> Ep.146

October 25th

The older brothers are bewildered.

형들둥절

hyeong-deul-dung-jeol

While strolling through the Korean Folk Village in a traditional outfit, V starts to recite a *sijo* (traditional Korean poem). Jin and SUGA are confused by the scene! The subtitle says, "형들둥절 hyeong-deul-dung-jeol," which means "The older brothers are bewildered." This is a phrase combining "형들 hyeong-deul" (older brothers) and "둥절 dung-jeol," which comes from "어리둥절 eo-ri-dung-jeol," meaning "being puzzled." When someone is bewildered, you can say "(the person being bewildered) + 둥절 dung-jeol."

03

Agreed

<Run BTS!> Ep.130

March 5th

agreed
인정
in-jeong

When Jin wins the final tennis match, V says Jin owes the victory to him, and the subtitle "**인정** in-jeong" appears above Jin, who agrees that this is true. "**인정** in-jeong" comes from the verb "**인정하다** in-jeong-ha-da" (to admit). When you agree with close friends in a casual conversation, you can simply say "**인정** in-jeong." When sending a text message to close friends, people sometimes type only the first consonants of Korean words, such as "ㅇㅈ." Did your friend say that BTS is gorgeous? Then you can reply "**인정** in-jeong!"

10

26

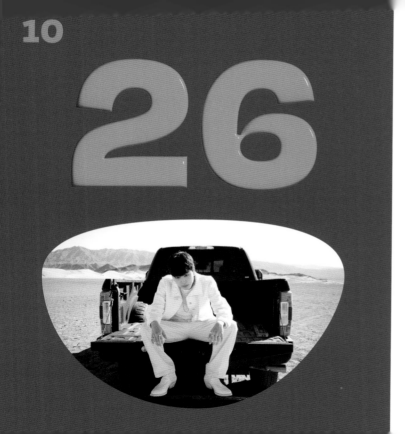

a binge-watching

정주행

jeong-ju-haeng

BTS is on a live video stream! Jimin says he is watching a certain entertainment program these days, using the expression "정주행 jeong-ju-haeng." While this expression originally means that a vehicle is going straight on a track, it can also refer to the act of watching a television series from the first episode to the last without taking a break. If you are binge-watching <Run BTS!> from the first episode, you can describe this situation by saying "정주행 jeong-ju-haeng."

03

Yeah, that's right. /
Why is that? Don't you just want to ride again?

<Run BTS!> Ep.154

yes yes, no no

ㅇㅇ, ㄴㄴ

Have you ever noticed "ㅇㅇ" or "ㄴㄴ" in the subtitles of <Run BTS!>, when BTS agrees or disagrees with something? "ㅇㅇ" is the initial consonants of "응응 eung-eung," which means "yes, yes," while "ㄴㄴ" is the initial consonants of "노노 no-no," the Korean spelling of "no, no." Both expressions are informal, so you should only use them in casual situations, like when texting your closest friends.

10

27

I was spacing out...

<Run BTS!> Ep.139

October 27th

I was spacing out.

멍 때렸어요.

meong ttae-ryeo-sseo-yo

While playing table tennis with V, Jimin zones out and misses the ball! He explains this by saying "멍 때렸어요 meong ttae-ryeo-sseo-yo," which means "I was spacing out." "멍 때리다 meong ttae-ri-da" is a combination of "멍하다 meong-ha-da" (to be in a daze) and "낮잠을 때리다 nat-ja-meul ttae-ri-da," a very casual way of saying "to take a nap." This new expression refers to an unresponsive state, as if the person is out of their mind. Do you ever go blank while doing something? Then try saying this expression: 멍 때렸어 meong ttae-ryeo-sseo, 멍 때렸어요 meong ttae-ryeo-sseo-yo.

03

03

Jin's team jinx

<Run BTS!> Ep.148

March 3rd

jinx

찌찌뽕

jji-jji-ppong

While listening to a question in a quiz, Jin and RM simultaneously shout their team name to indicate they want to give the answer. The subtitles describe this situation as "**찌찌뽕**jji-jji-ppong." This expression is used when two people say the same thing at the same time by chance. People usually shout it for fun. In Korea, the person who says "**찌찌뽕**jji-jji-ppong" pinches the other person's arm until the pinched person says something that has been specified, such as "**뽕찌찌**ppong-jji-jji." When you and your friend accidentally say the same thing at the same time, try saying "**찌찌뽕**jji-jji-ppong."

10

28

<Weverse Live> 2021.05.21

October 28th

Please pay attention!

�관부!

man-gwan-bu

During a live video stream celebrating their new release, BTS ends by saying "맘관부 man-gwan-bu" for their new song, <Butter>. This expression is an abbreviation using the first syllable of "많은 관심 부탁해요 ma-neun gwan-shim bu-ta-kae-yo," which means "Please pay attention" or "Please stay tuned." It is a new expression frequently used when promoting or advertising something. When BTS releases their next album, try saying "맘관부 man-gwan-bu!" to the people around you.

03

Our songs are so good.

<Run BTS!> Ep.144

so much

neom-na

BTS is dancing excitedly to their songs and the subtitles say "**넘나 좋은 우리 노래** neom-na jo-eun u-ri no-rae," which means "Our songs are so good." "**넘나** neom-na" is a casually used new term that is short for "**너무나** neo-mu-na," which means "so much." This is usually followed by an adjective. For example, you can use the adjective "**멋지다** meot-ji-da" (to be cool) and say "**넘나 멋져요** neom-na meot-jeo-yo!" to BTS.

10

29

<Run BTS!> Ep.143

H: 신박해요.
shin-ba-kae-yo

That's crazy.

신박하다.

shin-ba-ka-da

While listening to Jin's fairy tale, j-hope is amazed by the creative and surprising story. He says, "신박하다 shin-ba-ka-da," which is a new expression that means "to be new and crazy." This is commonly used by the younger generation instead of "신기하다 shin-gi-ha-da" (☛ July 14th), which means "to be amazing." If BTS releases an album with a style and concept you have never seen before, you can say "신박하다 shin-ba-ka-da!" or "신박해요 shin-ba-kae-yo!" in admiration.

03

01

놀이와 힐링, 그리고 일
그 사이의 달려라 방탄

Playing, healing, and working,
and <Run BTS!> somewhere in between

<Run BTS!> Ep.155

March 1st

healing

힐링

hil-ling

Jung Kook describes the experience of filming <Run BTS!> as a special time of "**놀이** no-ri, **힐링** hil-ling, **일** il," which means "playing," "healing," and "working," respectively. This implies that he felt healed as he had fun and relaxed while working. "**힐링** hil-ling" is the Korean spelling of "healing" in English. When BTS music brings you comfort, try using the expression "**힐링돼** hil-ling-dwae," or "**힐링돼요** hil-ling-dwae-yo."

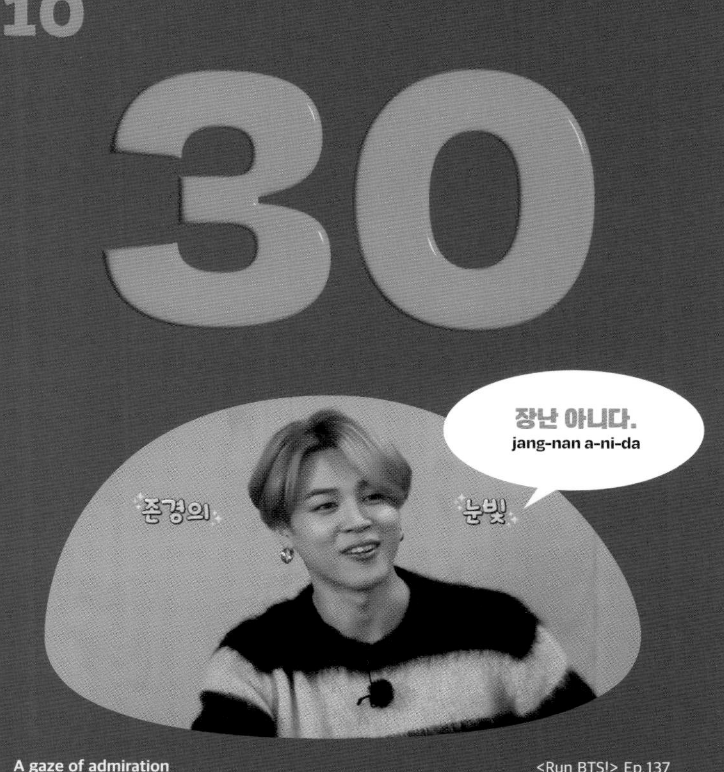

A gaze of admiration

<Run BTS!> Ep.137

H: 장난 아니에요.
jang-nan a-ni-e-yo

It's unbelievable.

장난 아니다.

jang-nan a-ni-da

BTS is playing a game of charades using the lyrics of their songs! j-hope starts to act out the lyrics of <DNA>. He uses his arms to make a spiral shape like DNA, and Jimin shouts "**장난 아니다** jang-nan a-ni-da!" to commend j-hope's sense of humor. This expression literally means that **something is not a joke** (**장난** jang-nan) and is used **when someone displays a perfect performance**. If you have witnessed something amazing or cool, try using "**장난 아니다** jang-nan a-ni-da!" or "**장난 아니에요** jang-nan a-ni-e-yo!" to express your awe and admiration.

3월
sa-mwol
March

10

31

리 액 션 맛 집

Reacting favorably

<Run BTS!> Ep.151

reacting favorably

리액션 맛집

ri-aek-shyeon mat-jip

Jung Kook reacts with various interjections, such as "Um" and "Oh," while watching TV! Reacting favorably to something like this can be described using the new expression "리액션 맛집 ri-aek-shyeon mat-jip." "맛집 mat-jip" refers to a popular restaurant that serves tasty food, but its meaning has been extended to include anything that has outstanding quality. So, "맛집 mat-jip" can be added after characteristics. For example, if there is a place where you can take beautiful selfies (셀카 sel-ka), it can be called "셀카 맛집 sel-ka mat-jip."

02

29

February 29th

H: 큰일났어요.
keu-nil-la-sseo-yo

We're in trouble.

keu-nil-la-sseo

During a tennis match in an episode of <Run BTS!>, j-hope expresses concern about his team being behind by two points, saying "큰일났어 keu-nil-la-sseo," which means "We're in trouble." "큰일 keu-nil" literally means "a huge incident," but in this expression, it means "a difficult and troublesome task" instead. In this case, j-hope is implying that it is hard to overcome two points. If you find that you don't have your wallet or phone when paying at a grocery store, you can say "큰일났어 keu-nil-la-sseo" or "큰일났어요 keu-nil-la-sseo-yo."

11월
shi-bi-rwol
November

02

28

RM Hyung! Please help me!

<Run BTS!> Ep.118

February 28th

Please help me.

도와주세요.

do-wa-ju-se-yo

Jung Kook, on a mission to take some photos during an episode of <Run BTS!>, suddenly says, "도와주세요 do-wa-ju-se-yo!" to ask for help from RM. This expression is used to ask for help in an urgent situation. You can casually say "도와줘 do-wa-jwo" to your close friends. If you need some help during your stay in Korea, try saying "도와주세요 do-wa-ju-se-yo."

11

01

It can't be Ttukseom Island! Really! Please!

<Run BTS!> Special Episode
- Telepathy Part 2

November 1st

please

제발

je-bal

BTS is supposed to gather at Hangang Park, but RM is worried that some of the others will go to the opposite side of the park. He desperately says, "제발 je-bal," which means "please," hoping that they will not. When you earnestly long for something or are asking a favor from somebody, you can use this expression. If you want BTS to come closer to your seat during a concert, show your eagerness by shouting "제발 je-bal!"

02

27

<Run BTS!> Ep.92

February 27th

H: 슬퍼요.
seul-peo-yo

It's sad.

seul-peu-da

j-hope fails to earn any points in a true-or-false quiz and says, "슬프다 seul-peu-da," which means "It's sad." You can use this expression with the Korean vowels "ㅠㅠ" or "ㅜㅜ," which look like crying eyes. You might feel sad when the BTS concert you have enjoyed a lot draws to an end. In that case, try using the expression "슬프다 seul-peu-daㅠㅠ" or "슬퍼요 seul-peo-yoㅜㅜ."

You bet.

<Run BTS!> Special Episode
- 'RUN BTS TV' On-air Part 1

H: 당연하죠.
dang-yeon-ha-jo

You bet.
당연하지.
dang-yeon-ha-ji

When Jung Kook says it is more comfortable being a viewer than a host on a live video stream, RM replies, "당연하지 dang-yeon-ha-ji." This expression can be used when a person's words are undoubtedly true. Some Koreans say, "당근이지 dang-geu-ni-ji," which is a conjugation of "당근 dang-geun," similar in pronunciation to "당연 dang-yeon." As "당근 dang-geun" means "carrot," the phrase is sometimes replaced with a carrot emoji🥕. When someone asks if you like BTS, answer like this: 당연하지 dang-yeon-ha-ji! 당연하죠 dang-yeon-ha-jo!

Cool, Chef SUGA.

<Run BTS!> Ep.103

H: 답답해요.
dap-da-pae-yo

I'm frustrated.
답답해.
dap-da-pae

RM and Jimin are cooking by following SUGA's orders. SUGA gets a chance to cook in person for two minutes. He quickly chops some onions and says, "답답해 dap-da-pae," expressing his frustration at being unable to cook as he intended. When something is beyond your control or someone is willing but unable to do something, you can use it. Try saying "답답해 dap-da-pae" or "답답해요 dap-da-pae-yo" when you want to say something in Korean, but nothing comes to mind.

03

<Run BTS!> Ep.95

You know what?

그거 알아요?

geu-geo a-ra-yo

Jimin uses the expression "그거 알아요 geu-geo a-ra-yo?" before talking to j-hope. This is used when sharing an interesting fact or emphasizing something important. You can casually say "그거 알아 geu-geo a-ra?" to your close friends. If you have figured out something interesting about BTS, try using this expression: 그거 알아 geu-geo a-ra? 그거 알아요 geu-geo a-ra-yo?

02

25

Ice-breaking master

<Run BTS!> Ep.147

○○ master

○○ 장인

jang-in

In an episode of <Run BTS!>, BTS members talk about Jin and describe him as someone who is not shy and is good at breaking the ice. The subtitles describe him as "ice-breaking 장인 jang-in." "장인 jang-in" refers to someone who is experienced in and excels at something. When someone is good at something, you can describe them as "○○ 장인 jang-in," putting the specialty in the position of "○○." For example, when someone is a good cook, you can say "요리 장인 yo-ri jang-in" by conjugating "요리 yo-ri," which means "cooking."

11

04

Is it over?

<Run BTS!> Special Episode
- 'RUN BTS TV' On-air Part 1

H: 끝난 거예요?
kkeun-nan geo-ye-yo

Is it over?
끝난 거야?
kkeun-nan geo-ya

Jin fails at the bear-feeding game! Jimin asks if that is the end of the game by saying "끝난 거야 kkeun-nan geo-ya?" This is an expression using "끝나다 kkeun-na-da" (to end). It is used when asking if something is finished, or to express regret about it. If you feel sad because the fun episode of <Run BTS!> is over, try saying "끝난 거야 kkeun-nan geo-ya?" or "끝난 거예요 kkeun-nan geo-ye-yo?"

02

It's crazy.

<Run BTS!> Special Episode
- Next Top Genius Part 1

February 24th

H: 미쳤어요.
mi-cheo-sseo-yo

It's crazy.

mi-cheot-da

BTS is playing a card game! Only Jin, RM, and Jimin have survived while the others have been eliminated. The winner must have a card marked "0," and all three have one! Jung Kook is surprised at this unexpected and dramatic situation, so he says, "미쳤다 mi-cheot-da." This expression literally means "to be crazy," but it is also used as an exclamation for something overwhelmingly wonderful or extraordinary. However, please note that this expression should only be used in comfortable situations among close friends. Have you already become proficient in Korean? 미쳤다 mi-cheot-da! 미쳤어요 mi-cheo-sseo-yo!

Oops.

<Run BTS!> Ep.110

Oops.

아이고 아이고야.

a-i-go a-i-go-ya

Jung Kook keeps a tissue off the ground for as long as 16 seconds by blowing on it! At the final moment, he says "아이고 아이고야 a-i-go a-i-go-ya," nearly losing his balance while trying to prevent the tissue from falling down. "아이고(야) a-i-go(ya)" is an exclamation that can be used in almost any situation, including when making a mistake or encountering a pleasant surprise. It is kind of a habit for Koreans to say it in those situations. You can use "에구 e-gu" or "에고 e-go" instead, which have the same meaning.

02

23

It's out, it's out, a legend is out. <Run BTS!> Ep.155

February 23rd

legend
레전드
re-jeon-deu

BTS must jump into the air and act out a given keyword with their bodies. As they do, a camera high in the air will take a picture of them. While many hilarious photos have been taken, SUGA creates the funniest shot for the keyword "hello" (안녕 an-nyeong 👉 August 2nd). Afterwards, the sentence "레전드 나왔다 re-jeon-deu na-wat-da" appears on the screen. "레전드 re-jeon-deu" is the Korean spelling of "legend" in English. When BTS gives an overwhelming and legendary performance worth rewatching, try using the expression: 레전드 re-jeon-deu!

11

Actually, I...

<Run BTS!> Ep.120

actually

사실

sa-shil

During a game, Jimin has to reveal his alibi to the others and begins with the expression "사실 sa-shil." This expression is used to state that you are about to disclose the truth in an honest manner. Do you have any secrets that you want to tell your friends? Then use the expression "사실 sa-shil" before you confess them.

02

22

Wow, you nailed <Run BTS!>!

<Run BTS!> Ep.128

February 22nd

H: 찢었어요.
jji-jeo-sseo-yo

You nailed it.

찢었다.

jji-jeot-da

While playing the Break the Eggs game with Jimin, RM smashes an egg with his palm! The egg's residue splatters all over Jimin's clothes, and everyone bursts into laughter. At that moment, the subtitle "찢었다 jji-jeot-da" appears. This is a slang used to express that something is amazing or awesome. In this case, RM's action of making all the others laugh just by breaking an egg is considered awesome, which is why the subtitle "찢었다 jji-jeot-da" is used. Let's also use the expression "찢었다 jji-jeot-da" or "찢었어요 jji-jeo-sseo-yo" when a BTS performance is so awesome!

11

Back to “Reading the instructions carefully” mode

<Run BTS!> Ep.149

I knew it.

어쩐지.

eo-jjeon-ji

While assembling furniture, Jin notices that one part has been inserted incorrectly. He says he knew something was weird and adds, “**어쩐지** eo-jjeon-ji.” This expression can be used when something you anticipated actually happens. For example, if you think the cloudy weather means it will rain soon and then it does, you can say, “**어쩐지** eo-jjeon-ji!”

02

21

가구 조립&페인팅 등
웬만한 셀프 인테리어 가능한 능력자들

People who are capable of doing self-interior work, such as furniture assembly and painting, etc

<Run BTS!> Ep.148

a person of skill

능력자

neung-nyeok-ja

When BTS says they have designed the interior of their own home, the subtitles read "능력자 neung-nyeok-ja." This refers to someone skilled or competent. BTS can be called "능력자 neung-nyeok-ja," as they sing difficult songs with powerful choreography. Do you have a friend with versatile talents? If so, you can call them "능력자 neung-nyeok-ja."

11

08

Oh, my goodness.

<Run BTS!> Ep.144

November 8th

Oh, my goodness!
이럴 수가!
i-reol ssu-ga

BTS has to guess their best performance, as chosen by ARMY! Everyone thinks hard and writes an answer. However, they're astonished when they hear that nobody gets the correct answer. The phrase "**이럴 수가** i-reol ssu-ga" appears on the screen, along with BTS' surprised faces. This is an exclamation used when you're shocked by something unbelievable. Is your holiday already over? **이럴 수가** i-reol ssu-ga!

02

20

He is also called "top-tier j-hope."

<Run BTS!> Ep.141

February 20th

top-tier ○○
갓○○
gat

BTS is spending some time praising each other. As j-hope is said to be a good example, the others call him "갓이홉 gat-i-hop." It is a combination of "갓 gat," the Korean spelling of "god," and "이홉 i-hop," from "제이홉 je-i-hop," the Korean spelling of "j-hope." Adding "갓 gat" before a noun means that someone or something is as outstanding as a god. Call your friend who gets front-row seats for every BTS concert "갓 gat + (the friend's name)."

11

09

November 9th

by any chance

hok-shi

While filming <Run BTS!>, V imagines himself talking to a student passing by outside the car window and asks himself, "혹시 아미세요 hok-shi a-mi-se-yo?" which means "Are you ARMY, by any chance?" "혹시 hok-shi" is an expression used when hesitating to say something you aren't 100% sure of. You may know that "hoxy" often appears in the subtitles on <Run BTS!>. This is an English pun of "혹시 hok-shi," using the same pronunciation. If you see someone looking at photos or watching videos of BTS at a café, you can ask them, "혹시 아미세요 hok-shi a-mi-se-yo?"

Shaking their heads

<Run BTS!> Ep.138

shaking one's head

do-ri-do-ri

BTS is watching a table tennis demonstration by two instructors. Their heads move left and right while watching the quick rally. At that moment, "도리도리 do-ri-do-ri" appears on the screen. The dictionary definition of this expression is "the movement of babies shaking their heads from side to side." It implies that BTS looks cute when moving their heads to follow the ball with their eyes. When adorable kids move their heads from left to right, try using the expression "도리도리 do-ri-do-ri!"

November 10th

Just because.

그냥요.

geu-nyang-yo

In an episode of <Run BTS!>, BTS is presented with a true-or-false quiz in which they must choose either "O" or "X" as the answer. When RM asks Jung Kook why he picked "X," he replies, "그냥요 geu-nyang-yo." This expression is used when there is no special reason for something. You can casually say "그냥 geu-nyang" to your close friends. If someone asks why a particular dish is your favorite, you can answer "그냥 geu-nyang" or "그냥요 geu-nyang-yo," unless there is a specific reason. However, be careful, as it might come across as insincere because it is a short answer without an explanation.

02

February 18th

18

Happy Birthday, j-hope!

생일 축하해요, 제이홉!

saeng-il chu-ka-hae-yo, j-hope

11

설마!

No way!

<Run BTS!> Special Episode
- Telepathy Part 2

No way!

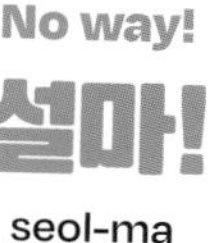

seol-ma

BTS is tasked with gathering at a place with sentimental value for all of them. V can't believe he is the only one at the place he chose and expresses his disbelief by saying, "설마 seol-ma." This is used when someone refuses to believe that something is true. For example, if the weather forecast says it will rain on the day of an outdoor BTS concert, you can say "설마 seol-ma" to express the hope that it isn't true.

02

17

<Run BTS!> Ep.144

We are the '95s, and a community of destiny.

a community of destiny
운명 공동체
un-myeong gong-dong-che

BTS is asked to guess their second-best healing song chosen by ARMY! When Jimin and V pick <Mikrokosmos>, "운명 공동체 un-myeong gong-dong-che" appears in the subtitles. "운명 un-myeong" means "destiny," and "공동체 gong-dong-che" means "community." So, "운명 공동체 un-myeong gong-dong-che" refers to people who face a common destiny, such as Jimin and V. They made the same choice and will share the same fate, whether they are right or wrong. BTS and ARMY can also be called "운명 공동체 un-myeong gong-dong-che," as they support and appreciate each other, right?

<Run BTS!> Special Episode
- Telepathy Part 1

November 12th

H: 그럼요!
geu-reom-yo

Of course!

그럼!

geu-reom

BTS is about to start a telepathy game. When RM asks if everyone is on the same page, Jung Kook answers, "그럼 geu-reom!" This implies that something is so obvious that there is no need to say it. It is commonly used to affirm that something is already known or agreed upon. If someone asks if you like BTS, you can say, "그럼 geu-reom!" or "그럼요 geu-reom-yo!"

02

16

Sparkling

<Run BTS!> Special Episode
- Next Top Genius Part 1

February 16th

sparkling

초롱초롱

cho-rong-cho-rong

BTS is playing the third round of the Liar Game! As V was the liar in the second round, his eyes are now sparkling brightly. The subtitles call them "초롱초롱 cho-rong-cho-rong," an expression that describes clear and vivid eyes. As V is no longer the liar in this round, he can enjoy himself and focus on the game. You can describe the shining eyes of BTS and ARMY when they see each other as "초롱초롱 cho-rong-cho-rong!"

11

13

Hey, Jung Kook. If it isn't 5...

<Run BTS!> Special Episode
- Next Top Genius Part 2

November 13th

if

만약에

ma-nya-ge

BTS is playing a dice game. When SUGA explains a situation in which a player can win with a particular number, he uses the expression "만약에 ma-nya-ge." This phrase is used at the beginning of a sentence to introduce a possible situation or condition that hasn't happened yet but might. Have you ever asked hypothetical questions like "If you met BTS while walking on the street" or "If you received comments from BTS on Weverse?" With those kinds of questions, you can use the expression "만약에 ma-nya-ge."

02

15

끄덕
끄덕
안경을 다시 쓰자 터지는 물 폭탄

Nodding /
The water bomb explodes when he puts the goggles back on.

<Run BTS!> Ep.132

February 15th

nodding
끄덕끄덕
kkeu-deok-kkeu-deok

BTS is playing a game where they have to avoid a particular action, and if they fail, they get hit with a water bomb! Jung Kook guesses that his forbidden action is putting on the goggles and deliberately does it. After he gets hit with a water bomb again, he nods, confirming his guess. The subtitles describe him as "끄덕끄덕 kkeu-deok-kkeu-deok," which is used to describe someone nodding. When someone asks if you want to go to a BTS concert, nod as vigorously as you want to. 끄덕끄덕 kkeu-deok-kkeu-deok!

11

14

<Run BTS!> Ep.151

November 14th

H: 이거예요!
i-geo-ye-yo

This is it!

이거야!

i-geo-ya

Jung Kook loves the room-service food he ordered from his hotel room and exclaims, "**이거야** i-geo-ya!" When you're more than satisfied with something, you can say this expression. Sometimes it can be said as "**바로 이거야** ba-ro i-geo-ya!" by adding "**바로** ba-ro," which means "exactly." When the latest BTS song is exactly the style you like best, you can say "**(바로) 이거야** (ba-ro) i-geo-ya!" or "**(바로) 이거예요** (ba-ro) i-geo-ye-yo!"

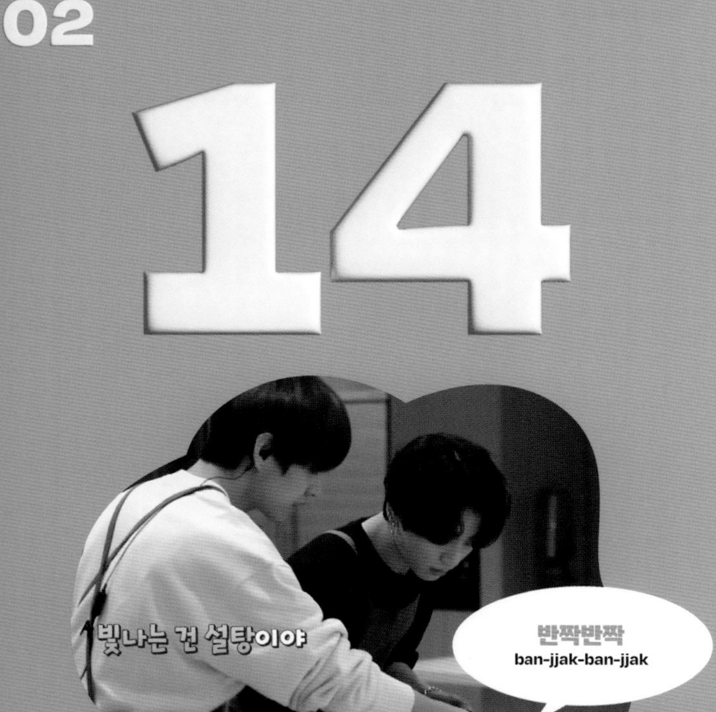

Sugar is the one that glitters.

<Run BTS!> Ep.102

February 14th

twinkle twinkle

반짝반짝

ban-jjak-ban-jjak

Can you distinguish salt from sugar with your bare eyes? V and Jung Kook cannot make the distinction, so Jin gives them a tip through a walkie-talkie, saying that sugar shines like "반짝반짝 ban-jjak-ban-jjak." This refers to something shiny that glitters. Your eyes are sparkling when you study Korean! 반짝반짝 ban-jjak-ban-jjak!

honestly
솔직히
sol-jji-ki

BTS is tasked with accurately recalling the lyrics of a song they have just listened to and then singing it in an episode of <Run BTS!>. When Jung Kook asks if anyone remembers the lyrics, RM and j-hope both use the phrase "**솔직히** sol-jji-ki" before admitting they don't remember. This expression, which means "honestly," is frequently used by Koreans to express their innocence or emphasize their opinion. You can start a conversation with "**솔직히** sol-jji-ki" when offering your candid opinions.

02

13

[BANGTAN BOMB]
2023 New Year Greetings from BTS

February 13th

I miss you.

보고 싶어요.

bo-go shi-peo-yo

While giving his New Year's greetings to ARMY, V says, "보고 싶어요 bo-go shi-peo-yo!" This expression literally means "I want to see you." If you love or miss someone, you naturally want to see them all the time, right? "보고 싶어요 bo-go shi-peo-yo" can be used when you miss someone a lot. Did you know that BTS song <Spring Day> starts with the lyrics "보고 싶다 bo-go ship-da?" Whenever you miss BTS, try using the expression "보고 싶어 bo-go shi-peo" or "보고 싶어요 bo-go shi-peo-yo."

11

16

Me too. (Me neither.)

저도요.

jeo-do-yo

In an episode of <Run BTS>, when RM admits that he doesn't know how to play the harmonica, Jin responds with "**저도요** jeo-do-yo," implying that he can't play the instrument either. This expression means "Me too" or "Me neither." It is often used to react to a statement and convey that you're in the same situation. You can casually say "**나도** na-do" to your close friends. If someone says they like BTS, try responding with "**나도** na-do!" or "**저도요** jeo-do-yo!" to show that you like BTS too.

02

12

<Weverse Live> 2020.05.29

It gives me a lot of strength.

힘을 많이 얻어요.

hi-meul ma-ni eo-deo-yo

On a live video stream after the release of his new solo mixtape "D-2," SUGA says that making an album gives him a lot of strength. In this situation, you can say "힘을 많이 얻어요 hi-meul ma-ni eo-deo-yo," which means "It gives me a lot of strength," like SUGA does. Do you feel energized when you watch BTS? Then try using this expression: 힘을 많이 얻어 hi-meul ma-ni eo-deo, 힘을 많이 얻어요 hi-meul ma-ni eo-deo-yo.

<Run BTS!> Ep.132

What should I say?

mwo-ral-kka

BTS is discussing bread with raisins, and when j-hope tries to explain why he doesn't like it, he begins his sentence with the expression "뭐랄까 mwo-ral-kka." This expression is used when someone is hesitant to answer a question, unsure how to express something, or doesn't know what to say. If you're asked why you love BTS songs but don't know how to answer, you can use the expression "뭐랄까 mwo-ral-kka" as a way to start your response.

02

February 11th

I received comfort and support.

위로와 응원을 받았어요.

wi-ro-wa eung-wo-neul ba-da-sseo-yo

On a live video stream, V explains that he received strength from ARMY's posts on Weverse when he couldn't record a song because of a cold. He says, "위로와 응원을 받았어요 wi-ro-wa eung-wo-neul ba-da-sseo-yo." The word "위로 wi-ro," which means "comfort," is also included in their song that comforts ARMY, <Magic Shop>. "응원 eung-won" means "support," and "받다 bat-da" means "to receive." So "위로와 응원을 받았어요 wi-ro-wa eung-wo-neul ba-da-sseo-yo" implies that you have received comfort and support. You might be consoled and encouraged by BTS music. Why don't you tell them this: 위로와 응원을 받았어요 wi-ro-wa eung-wo-neul ba-da-sseo-yo?

18

Just give me anything!

<Run BTS!> Ep.96

anything
아무거나
a-mu-geo-na

BTS is about to play with some tops, and everyone is excited to choose their own tops except for SUGA, who says "**아무거나** a-mu-geo-na," implying that anything is fine. This expression is used when the choice doesn't matter. For example, if you're picking a <Run BTS!> episode to watch with your friends at home, and you're okay with anything because all the episodes are funny, you can say "**아무거나** a-mu-geo-na!"

02

10

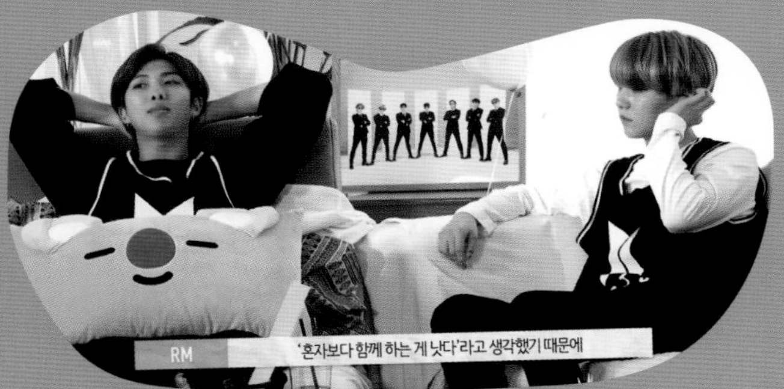

RM
'혼자보다 함께 하는 게 낫다'라고 생각했기 때문에

[2020 FESTA] BTS (방탄소년단) '방탄생파'

Because I thought it was better

February 10th

It's better together than alone.

혼자보다 함께 하는 게 낫다.

hon-ja-bo-da ham-kke ha-neun ge nat-da

RM believes that BTS has been successful so far because they have worked together rather than alone. "혼자보다 함께 하는 게 낫다 hon-ja-bo-da ham-kke ha-neun ge nat-da" is how you would say it in Korean. "혼자 hon-ja" means "alone," and "함께 ham-kke" means "together." This expression is a good reflection of Korean culture, which tends to value the collective rather than the individual. In difficult times, it's better to stay together and help each other out than to be alone, don't you think?

11

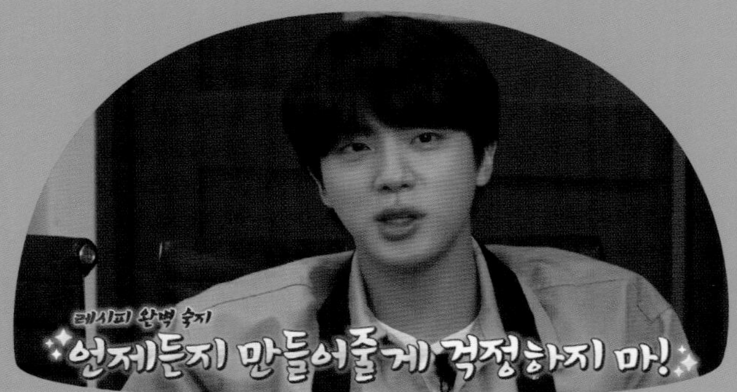

Perfect mastery of the recipe /
I can cook for you anytime. Don't worry!

<Run BTS!> Ep.122

November 19th

anytime

언제든지

eon-je-deun-ji

V is enjoying the food that Jin cooked, and Jin says he's happy to make it for V anytime. He uses the expression "언제든지 eon-je-deun-ji," which means "always" or "anytime." It's a combination of "언제 eon-je," which indicates a point in time that's not fixed, and "-든지 deun-ji," which means any choice is fine. If your friends ask if they can visit your house again after having fun together, you can say "언제든지 eon-je-deun-ji!" to let them know they're always welcome.

ARMY Forever, BTS Forever

아포방포

a-po-bang-po

While Jung Kook was chatting online with ARMY, several buzzwords were made. One of them is "아포방포 a-po-bang-po." This is an expression created by taking the first letter of each word in "아미 포에버 방탄 포에버 a-mi po-e-beo bang-tan po-e-beo," which is the Korean pronunciation of "ARMY Forever, BTS Forever" in English. 아포방포 a-po-bang-po!

11

20

RM
달방 덕분에 예능에 대한 그런 게 많이 없어졌죠

My fear of appearing on entertainment shows has disappeared a lot thanks to <Run BTS!>.

<Run BTS!> Ep.155

November 20th

thanks to it

덕분에

deok-bu-ne

RM says that his fear of appearing on entertainment shows has disappeared a lot thanks to <Run BTS!>. He uses the expression "**덕분에** deok-bu-ne." This phrase is used to attribute the cause of a positive event or result, using the structure "N (noun) **덕분에** deok-bu-ne." If you have ever been encouraged by the heartwarming and comforting lyrics of BTS songs, you can say "**방탄소년단 덕분에** bang-tan-so-nyeon-dan deok-bu-ne, I gained strength!" to acknowledge the positive influence that BTS has had on you.

02

08

Jung Kook: Jimin Hyung! I'm on your side now.

<Run BTS!> Ep.127

February 8th

H: 형의 편이에요.
hyeong-ui pyeo-ni-e-yo

I'm on your side.

형의 편이야.

hyeong-ui pyeo-ni-ya

Before a game of rock-paper-scissors, Jung Kook cheers Jimin on and says, "형의 편이야 hyeong-ui pyeo-ni-ya," meaning that he is on Jimin's side. To express that you're with someone, you can use the phrase "N (noun)의 편 ui pyeon," as Jung Kook does. You can omit "의 ui" in this expression and just say "N (noun) 편 pyeon." When a close friend is having a hard time, you can say "너의 편이야 neo-ui pyeo-ni-ya" or "네 편이야 ne pyeo-ni-ya," using "너 neo" or "네 ne," which means "you" and "your" respectively. Your friend will be very grateful for your encouragement.

11

21

<Run BTS!> Ep.127

Clear, right?

깔끔하지?

kkal-kkeum-ha-ji

After explaining his plan to complete the mission and make a closing remark, Jimin says, "깔끔하지 kkal-kkeum-ha-ji?" to express that his explanation was crystal clear. "깔끔하다 kkal-kkeum-ha-da" is an expression used to describe something that is clean and tidy. In honorifics, you can say "깔끔하죠 kkal-kkeum-ha-jo?" to sound more formal and polite. When giving a clear answer to someone's question, try using this expression: 깔끔하지 kkal-kkeum-ha-ji? 깔끔하죠 kkal-kkeum-ha-jo?

02

07

Such. A. Shame.

<Run BTS!> Ep.138

February 7th

H: 아쉬워요.
a-shwi-wo-yo

What a shame.

아쉽다.

a-shwip-da

Jung Kook is trying to hit a plastic bottle with a ping-pong ball. When the ball nearly hits the target, Jin says, "너무 아쉽다 neo-mu a-shwip-da." "아쉽다 a-shwip-da" is an expression that can be used when you miss out on something you wanted or have to leave a place that you've been enjoying. It's a way of expressing regret or disappointment. "너무 neo-mu" (too much) emphasizes the weight of the following word. If the team you are rooting for loses the game by a single point, try using the expression: 아쉽다 a-shwip-da! 아쉬워요 a-shwi-wo-yo!

22

I feel happy.

행복해요.

haeng-bo-kae-yo

After winning an award at a music award ceremony in the US in 2021, BTS turns on their live video stream. While speaking live, Jung Kook says he's happy, saying "**행복해요** haeng-bo-kae-yo." ARMY may already know this expression. In Jung Kook's heartfelt message to ARMY, "**아무행알** a-mu-haeng-al," which means "ARMY should be happy no matter what happens. Okay?" "**행** haeng" is the letter which comes from "**행복하다** haeng-bo-ka-da" (to be happy).

02

Jin: This is... more unexpected than you think.

<Run BTS!> Ep.154

February 6th

This is unexpected.

의외예요.

ui-oe-ye-yo

Jin announces each member's best scene from <Run BTS!> as selected by ARMY! BTS makes guesses, and Jin says the correct answer is unexpected before unveiling Jung Kook's result, saying "**의외예요** ui-oe-ye-yo." The expression "**의외** ui-oe" refers to something unexpected. You can casually say "**의외야** ui-oe-ya" to your close friends. If a quiet friend goes to a singing room and starts dancing and singing like BTS, it would be surprising. In that case, you can try saying "**의외야** ui-oe-ya," or "**의외예요** ui-oe-ye-yo."

11

23

<Run BTS!> Ep.142

November 23rd

What about ○○?

○○ 어때요?

eo-ttae-yo

BTS is discussing which dish they will make, and when Jin suggests *sujebi* (hand-pulled dough soup), he asks, "**수제비 어때요** su-je-bi eo-ttae-yo?" The expression "**○○ 어때요** eo-ttae-yo?" is used to suggest or recommend something. For instance, when you find clothes that look good on your friend while shopping, you can say "**이거 어때** i-geo eo-ttae?" or "**이거 어때요** i-geo eo-ttae-yo?" with the addition of "**이거** i-geo," which means "this."

02

05

<Run BTS!> Ep.140

February 5th

H: 헷갈렸어요.
het-gal-lyeo-sseo-yo

I was confused.

헷갈렸어.

het-gal-lyeo-sseo

While trying to guess who the celebrities in photos are, j-hope mistakenly shouts out the wrong name, even though he knows who the person is. He then says "헷갈렸어 het-gal-lyeo-sseo" several times to express his embarrassment. This expression is used when you were confused. If you ever find yourself giving the wrong answer because you were confused during a quiz or exam, try saying "헷갈렸어 het-gal-lyeo-sseo" or "헷갈렸어요 het-gal-lyeo-sseo-yo."

11

By the way...

<Run BTS!> Ep.129

by the way

아니 근데

a-ni geun-de

RM uses the expression "아니 근데 a-ni geun-de" to shift the conversation's focus and emphasize his next point. Although "아니 a-ni" is typically negative, Koreans sometimes use it at the start of a sentence out of habit to emphasize something. "근데 geun-de" is an abbreviated form of "그런데 geu-reon-de" and introduces a topic that is unrelated to or different from the current discussion. When you speak with your Korean friends, try using "아니 근데 a-ni geun-de" to begin a sentence like RM.

04

Jung Kook HBD

<Weverse Live> 2021.09.01

February 4th

H: 애매하네요.
ae-mae-ha-ne-yo

It's ambiguous.
애매하네.
ae-mae-ha-ne

Before finishing a live video stream, Jung Kook checks the time, which is 1:09 a.m. He then says, "애매하네 ae-mae-ha-ne," and asks ARMY if he should continue until 1:30 a.m. Jung Kook is not the only one who prefers starting or finishing something precisely, such as at 1:00 or 1:30. When something is insufficient, ambiguous, and totally unsatisfying, try using the expression "애매하네 ae-mae-ha-ne," or "애매하네요 ae-mae-ha-ne-yo."

11

25

<Run BTS!> Ep.149

November 25th

That's what I'm saying.

내 말이.

nae ma-ri

Jin explains the interior design of the room he created with RM and emphasizes that living in such a nice place is uncommon. RM agrees with Jin, saying "**내 말이** nae ma-ri!" This is short for "**내 말이 그 말이야** nae ma-ri geu ma-ri-ya," which means "That's what I'm saying." It is used when you strongly agree with what someone else said. For example, if your friend is talking about what makes BTS lovable and you feel the same way, you can say "**내 말이** nae ma-ri!"

02

03

I'm afraid of water.

<Run BTS!> Ep.124

February 3rd

H: 물이 무서워요.
mu-ri mu-seo-wo-yo

I'm afraid of water.

물이 무서워.

mu-ri mu-seo-wo

BTS is selecting a sport to be part of a long-term project of <Run BTS!>. When Jung Kook suggests swimming, j-hope says "**난 물이 무서워** nan mu-ri mu-seo-wo," which means "I'm afraid of water." "**무서워** mu-seo-wo" is an expression used when you are scared of specific situations or objects. Are you afraid of anything? If so, you can say "**무서워** mu-seo-wo" or "**무서워요** mu-seo-wo-yo."

11

26

This is much more difficult than I thought!

<Run BTS!> Ep.139

November 26th

○○ than I thought

생각보다 ○○

saeng-gak-bo-da

V is playing table tennis and finds it more difficult than he thought, using the expression "생각보다 saeng-gak-bo-da," which means "than I thought." This expression is commonly used before adjectives to indicate that a specific condition is more than expected. If you want to express that something is more challenging than expected, you can say "생각보다 어렵다 saeng-gak-bo-da eo-ryeop-da," using the adjective "어렵다 eo-ryeop-da" (to be difficult).

02

The two pairs born in the same year, '94 and '95, are hustling and bustling.

<Run BTS!> Ep.127

a person the same age as you

dong-gam-nae-gi

Did you know that BTS has two pairs of members born in the same year? RM and j-hope were born in 1994, while Jimin and V were born in 1995. When the four members are playing games, the subtitle "동갑내기 dong-gam-nae-gi" appears. "동갑 dong-gap" means "the same age," and "동갑내기 dong-gam-nae-gi" refers to a person born in the same year. Do you have any close friends who were born in the same year as you, just like RM and j-hope, and Jimin and V?

11

27

November 27th

H: 나쁘지 않은 시도였어요.
na-ppeu-ji a-neun shi-do-yeo-sseo-yo

It was a good try.

나쁘지 않은 시도였어.

na-ppeu-ji a-neun shi-do-yeo-sseo

In an episode of <Run BTS!>, Jin plays a game where he has to make a bear eat a cake. After watching him try many different ways, SUGA comments, "**나쁘지 않은 시도였어** na-ppeu-ji a-neun shi-do-yeo-sseo," meaning that Jin's attempt (**시도** shi-do) was not bad. This expression is used to evaluate an attempt that is not perfect but fairly close to success. You can say "**나쁘지 않은 시도였어(요)** na-ppeu-ji a-neun shi-do-yeo-sseo(yo)" when evaluating such an attempt.

02

한마음 한뜻
han-ma-eum han-tteut
모두가 응원하는
김치 칼국수 같은 수제비

Everyone cheers for *sujebi*, which was going to be *kimchi kalguksu* before.

<Run BTS!> Ep.123

February 1st

one mind and one goal

한마음 한뜻

han-ma-eum han-tteut

As Jin and SUGA make *kimchi sujebi* (hand-pulled dough soup with kimchi), RM says that everyone is cheering for them in unison and adds, "한마음 한뜻 han-ma-eum han-tteut." "한마음 han-ma-eum" means "one mind," and "한뜻 han-tteut" means "one goal." In other words, "한마음 한뜻 han-ma-eum han-tteut" implies that everyone is like-minded, just like ARMY when they support BTS!

11

28

<Run BTS!> Ep.92

H: 포기하지 마세요.
po-gi-ha-ji ma-se-yo

Don't give up.

포기하지 마.

po-gi-ha-ji ma

j-hope finally gets the right answer in a quiz after getting all the previous questions wrong. As the quiz's host, V cheers for j-hope by saying, "**포기하지 마** po-gi-ha-ji ma." "**포기하다** po-gi-ha-da" means "to give up." If you add "**-지 마** ji ma" after it, it means "not to do something." For the polite expression, you can say, "**-지 마세요** ji ma-se-yo." Are you finding it challenging to study Korean? You're doing great. **포기하지 마세요** po-gi-ha-ji ma-se-yo!

2월
i-wol
February

29

<Run BTS!> Ep.95

H: 가능해요.
ga-neung-hae-yo

It's possible.

가능해.

ga-neung-hae

BTS is playing a game by moving a bottle cap three times to conquer more territory! SUGA made two decent attempts to secure a good position, and his final attempt seems promising. RM says "**가능해** ga-neung-hae!" because it seems like SUGA has a chance to win the match. The expression "**가능해(요)** ga-neung-hae(yo)" is used when something is possible. On the other hand, when something is impossible, try saying "**불가능해(요)** bul-ga-neung-hae(yo)."

01

31

lol /
Are you sure you are just a bit nervous?

<Run BTS!> Ep.124

January 31st

H: 좀 떨리네요.
jom tteol-li-ne-yo

I'm a bit nervous.

좀 떨리네.

jom tteol-li-ne

Before a presentation, Jin says "좀 떨리네 jom tteol-li-ne." "떨리네 tteol-li-ne" is a conjugation of "떨리다 tteol-li-da," which means "to be shaky and nervous" and "좀 jom" means "a little bit." If the weather is a bit chilly, you can say "좀 춥네 jom chum-ne," conjugating "춥다 chup-da" (to be cold). Imagine a BTS concert is about to begin. If you feel nervous, you can say "떨리네 tteol-li-ne" or "떨리네요 tteol-li-ne-yo."

11

30

<Run BTS!> Ep.110

H: 좋은 생각이에요.
jo-eun saeng-ga-gi-e-yo

Great idea.

좋은 생각이다.

jo-eun saeng-ga-gi-da

BTS must stand on a shrinking piece of newspaper in a game! RM comes up with a good idea, and Jin says "좋은 생각이다 jo-eun saeng-ga-gi-da" to him. When you like someone else's idea (생각 saeng-gak), you can use this expression to agree with it. The phrase is usually used as an exclamation, and to make it more formal and polite, you can say "좋은 생각이에요 jo-eun saeng-ga-gi-e-yo." Do you plan to study Korean every day? 좋은 생각이에요 jo-eun saeng-ga-gi-e-yo!

01

30

<Weverse Live> 2021.12.03

January 30th

It's bittersweet.

시원섭섭해요.

shi-won-seop-seo-pae-yo

After a concert that BTS had spent a long time preparing for, RM says, "**시원섭섭해요** shi-won-seop-seo-pae-yo." If you have worked hard for a long time to prepare for something, you could have mixed feelings when it is over. While "**시원하다** shi-won-ha-da" describes a cold temperature, it is also used when you feel refreshed and cleansed. "**섭섭하다** seop-seo-pa-da" describes a sad and regretful feeling that arises when something ends or when you have to say goodbye to someone. Therefore, when you feel both refreshed and sad at the same time, you can express the bittersweet feeling by saying "**시원섭섭해** shi-won-seop-seo-pae" or "**시원섭섭해요** shi-won-seop-seo-pae-yo."

12월
shi-bi-wol
December
HAPPY HOLIDAY

01

Calmness itself

<Run BTS!> Special Episode
- Fly BTS Fly Part 1

itself

그 잡채

geu jap-chae

Jimin is peacefully doing flying yoga poses and "평온 그 잡채 pyeong-on geu jap-chae" appears in the subtitles. The correct expression is "평온 그 자체 pyeong-on geu ja-che," which consists of the noun "평온 pyeong-on" (calmness) followed by "그 자체 geu ja-che" (itself). Adding "그 자체 geu ja-che" after a noun highlights the distinct and remarkable characteristic of someone or something. As the word "자체 ja-che" and the Korean food "잡채 jap-chae" have similar pronunciations, the younger generation has popularized replacing "자체 ja-che" with "잡채 jap-chae." If you cheer for BTS with an ARMY BOMB in your hand, you can be called "ARMY 그 잡채 geu jap-chae!"

12

H: 긴장하지 마세요.
gin-jang-ha-ji ma-se-yo

Relax.

긴장하지 마.

gin-jang-ha-ji ma

Before a table tennis match between Jimin and V in an episode of <Run BTS!>, Jin says "**긴장하지 마** gin-jang-ha-ji ma" to V, who looks nervous. This expression can be used to encourage people you are cheering for to regain their composure and demonstrate their abilities. To make it more formal and polite, you can say "**긴장하지 마세요** gin-jang-ha-ji ma-se-yo." Try saying this to someone who is nervous before an exam, presentation, or performance: **긴장하지 마** gin-jang-ha-ji ma! **긴장하지 마세요** gin-jang-ha-ji ma-se-yo!

Precious

<Run BTS!> Ep.110

January 28th

precious

소듕

so-dyung

Jung Kook is carefully holding onto his treasure chest! At that moment, the word "소듕 so-dyung" shows up on the screen. Its original form is "소중 so-jung," which comes from the basic form "소중하다 so-jung-ha-da," which means "to be precious." However, in this case, the syllable "중 jung" is pronounced as "듕 dyung" to sound as if a cute child were saying it. As the image of Jung Kook holding onto the box is adorable, the subtitle is appropriate.

It's similar.

비슷해요.

bi-seu-tae-yo

In an episode of <Run BTS!>, V is acting as the host of a quiz show and says, "**비슷해요** bi-seu-tae-yo. **땡** ttaeng!" when Jimin says something close to the right answer but gets it wrong. You may already know that "**땡** ttaeng" (👉 June 19th) is an expression used by the host when an answer is incorrect. "**비슷해요** bi-seu-tae-yo" is used when two things are similar without much difference. You can casually say "**비슷해** bi-seu-tae" to your close friends. If someone gives a similar but incorrect answer to a quiz question, try saying "**비슷해** bi-seu-tae" or "**비슷해요** bi-seu-tae-yo."

A king of positivity

<Run BTS!> Ep.144

a king of positivity

긍정왕

geung-jeong-wang

BTS is playing a game where they have to guess song titles based on the Korean initials of some of the lyrics. Even though Jin keeps getting the answers wrong, he remains positive and says he will keep trying. The subtitle describes him as "**긍정왕** geung-jeong-wang," a compound word of "**긍정** geung-jeong" (positivity) and "**왕** wang" (king), which means a very positive person. Adding "**왕** wang" at the end of a noun implies that the person is highly skilled at something. Another example is "**요리왕** yo-ri-wang," which is a combination of "**요리** yo-ri" (cooking) and "**왕** wang."

12

03

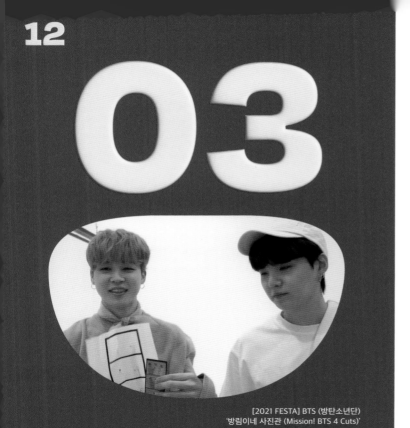

[2021 FESTA] BTS (방탄소년단)
'방림이네 사진관 (Mission! BTS 4 Cuts)'

December 3rd

It's the same.

똑같아요.

ttok-ga-ta-yo

Jimin compares four pictures with the photos he took by saying "똑같아요 ttok-ga-ta-yo." When more than two things are the same, such as a pair of chopsticks or socks, you can use this expression. What else is the same? The love that BTS and ARMY have for each other, 똑같아요 ttok-ga-ta-yo!

01

26

저랑 찰떡이에요.
jeo-rang chal-tteo-gi-e-yo
골드가 곧 나

Gold suits me.

<Run BTS!> Ep.104

January 26th

a perfect match

찰떡

chal-tteok

When the staff asks BTS if they need any additional items to decorate their clothes, Jung Kook requests a gold chain, saying "**저랑 찰떡이에요** jeo-rang chal-tteo-gi-e-yo." "**찰떡** chal-tteok" refers to a very sticky type of rice cake, so when two things or people are a perfect match, they can be described as "**찰떡** chal-tteok," meaning they stick well together. In that sense, BTS and ARMY are **찰떡이에요** chal-tteo-gi-e-yo!

12

December 4th

04

Happy Birthday, Jin!

생일 축하해요, 진!

saeng-il chu-ka-hae-yo, Jin

01

Ha-ha-ha

<Run BTS!> Ep.143

January 25th

ha-ha-ha

kkal-kkal-kkal

BTS is laughing out loud at the title and cover of the fairy tale that j-hope and V made! The subtitle "깔깔깔 kkal-kkal-kkal," which appears on the screen at that moment, is an expression that represents the sound of laughing heartily and cheerfully, unable to hold back. When you tell funny stories to your friends or can't stifle a laugh while watching <Run BTS!>, try this expression: 깔깔깔 kkal-kkal-kkal!

12

05

<Run BTS!> Ep.102

December 5th

pat pat

토닥토닥

to-dak-to-dak

V and Jung Kook are cooking by following Jin's instructions! When Jin orders Jung Kook to pat V on the bottom to show support, he uses the expression "**토닥토닥** to-dak-to-dak." This is used to describe the sound or action of patting something lightly. When BTS is about to call it a day, try saying "**토닥토닥** to-dak-to-dak" to tell them they did a good job.

01

Clapping

<Run BTS!> Ep.144

January 24th

clapping
짝짝
jjak-jjak

When BTS starts filming an episode of <Run BTS!>, they shout "달려라 방탄 dal-lyeo-ra bang-tan," which means "Run BTS," and clap their hands. At that moment, the subtitle "짝짝 jjak-jjak" appears along with an image of clapping hands 👏. This expression refers to the sound of clapping hands and is also included in one of BTS songs, <2! 3!>. Have you heard "박수 짝짝 bak-su jjak-jjak" in SUGA's rap? "박수 bak-su" means "clap." For everyone who has worked hard at studying Korean today, 짝짝 jjak-jjak!

12

06

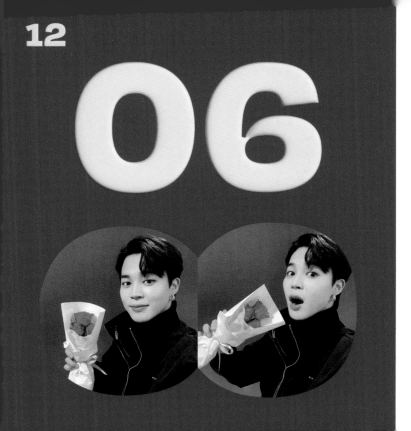

a gratitude

감사한 마음

gam-sa-han ma-eum

During an interview, Jimin says he is grateful to SUGA for helping him try something. This feeling of gratitude towards someone is called "**감사한 마음** gam-sa-han ma-eum," conjugating "**감사하다** gam-sa-ha-da," which means "to be grateful" and "**마음** ma-eum," which means "heart" or "mind." If you have an apology in your heart for someone, you can say that you have "**미안한 마음** mi-an-han ma-eum" by conjugating "**미안하다** mi-an-ha-da" (to be sorry). What kind of "**마음** ma-eum" do you have for BTS?

23

I'm anxious.

<Run BTS!> Ep.131

H: 불안한데요?
bu-ran-han-de-yo

I'm anxious.

불안한데?

bu-ran-han-de

BTS is filming <Run BTS!> at a swimming pool! j-hope asks what they are doing that day and says "불안한데 bu-ran-han-de?" "불안하다 bu-ran-ha-da" is used when expressing anxiety. In this case, j-hope is concerned that they might have to go in the pool and get wet. Do you feel anxious when you think you may have left something at home, or when you are about to give a presentation without being fully prepared? Try using the expression "불안한데 bu-ran-han-de?" or "불안한데요 bu-ran-han-de-yo?" in those situations.

12

07

Sufficiency

<Run BTS!> Ep.123

December 7th

H: 충분해요.
chung-bun-hae-yo

That's enough.
충분해.
chung-bun-hae

j-hope must complete a dish by giving orders to the avatar chefs, Jin and SUGA, by radio! While thinking of step-by-step strategies, he says "충분해 chung-bun-hae" to indicate that they have plenty of time. This expression can be used when there is enough of something, with no shortages. If someone asks you about your cell phone battery when it's 100% charged, you can answer like this: 충분해 chung-bun-hae, 충분해요 chung-bun-hae-yo.

01

22

j-hope: Whoa, I'm so shocked.

<Run BTS!> Special Episode
- Telepathy Part 1

January 22nd

H: 충격적이에요.
chung-gyeok-jeo-gi-e-yo

I'm shocked.

충격적이야.

chung-gyeok-jeo-gi-ya

BTS is on a mission to gather in the same location. j-hope hears that no one came to the location he chose and says "**충격적이야** chung-gyeok-jeo-gi-ya." This expression is used when you are in shock because of an unexpected or unbelievable situation. When BTS astonishes you with their unprecedented comebacks and unbelievable performances, you can say "**충격적이야** chung-gyeok-jeo-gi-ya!" or "**충격적이에요** chung-gyeok-jeo-gi-e-yo!" to express your shock.

12

08

<Run BTS!> Ep.142

December 8th

H: 단순하네요.
dan-sun-ha-ne-yo

It's simple.

단순하네.

dan-sun-ha-ne

BTS is learning to cook! j-hope thought it would be complicated, but he says "**단순하네** dan-sun-ha-ne" when the recipe turns out to be simpler than expected. "**단순** dan-sun" means "to be simple with no complexity." Conversely, when something is complicated, you can say "**복잡하네(요)** bok-ja-pa-ne(yo)." When someone around you is struggling with a problem that seems simple to you, propose a solution by using this expression: **단순하네(요)** dan-sun-ha-ne(yo)!

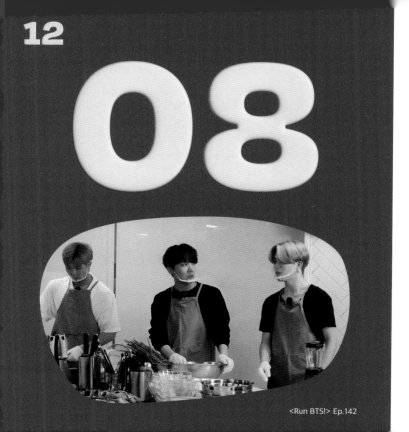

01

21

January 21st

H: 당황스러워요.
dang-hwang-seu-reo-wo-yo

I'm embarrassed.

당황스러워.

dang-hwang-seu-reo-wo

In a dancing quiz during an episode of <Run BTS!>, Jin keeps guessing the correct answers while the staff dances to BTS songs! As Jin did not expect to be so good at the quiz, he is embarrassed and says "당황스러워 dang-hwang-seu-reo-wo." You can say this expression when you encounter an unexpected situation and feel embarrassed. When someone asks you a difficult question, try saying "당황스러워 dang-hwang-seu-reo-wo" or "당황스러워요 dang-hwang-seu-reo-wo-yo" to express your embarrassment.

12

09

But it's you!

<Run BTS!> Ep.146

H: 당신밖에 없어요.
dang-shin-ba-kke eop-seo-yo

You're the only one.

너밖에 없어.

neo-ba-kke eop-seo

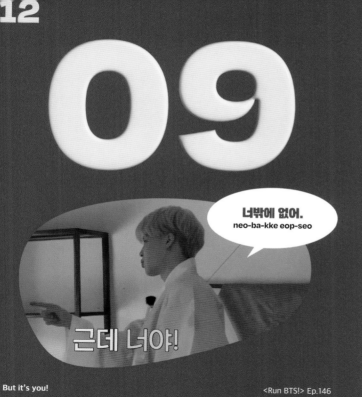

Jimin is trying to figure out who is playing the role of a thief. He suspects V and says, "**너밖에 없어** neo-ba-kke eop-seo." The phrase "**N** (noun)**밖에** ba-kke" means "except for **N** (noun)." So, Jimin is conveying that V is the only person he suspects of being the thief. It's usually followed by words with negative connotations, such as "**모르다** mo-reu-da" (do not know) or "**없다** eop-da" (do not exist). "**너밖에 없어** neo-ba-kke eop-seo" can also be used to express gratitude to a very dear and close friend.

01

20

Jealous?

<Run BTS!> Ep.151

January 20th

H: 부러워요.
bu-reo-wo-yo

I'm jealous.

bu-reop-da

BTS is waiting for their food to arrive, but their order is running late. Finally, Jung Kook's order arrives first, and RM says, "부럽다 bu-reop-da." This expression is used when you are jealous of someone who has or does what you want before you. In honorifics, "부러워요 bu-reo-wo-yo" can be said to make it sound formal and polite. Did you come across BTS during your visit to Korea? 부러워요 bu-reo-wo-yo!

12

10

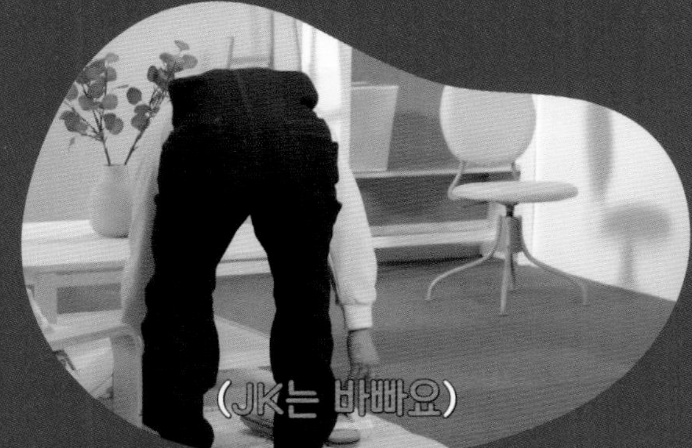

Jung Kook is busy.

<Run BTS!> Ep.149

December 10th

I'm busy.

바빠요.

ba-ppa-yo

SUGA and Jimin are about to move a table! They ask Jung Kook to help them, but he's busy doing something. At that moment, "**JK는 바빠요** neun ba-ppa-yo," which means "Jung Kook is busy," appears in the subtitles. "**바빠요** ba-ppa-yo" is an expression used when you're busy with a lot of work. You can casually say "**바빠** ba-ppa" to your close friends. When someone asks you something while you're busy watching BTS video clips, you can use this expression: **바빠** ba-ppa, **바빠요** ba-ppa-yo.

01

BTS (방탄소년단) 'BE' Comeback Countdown

January 19th

I don't know.

잘 모르겠어요.

jal mo-reu-ge-sseo-yo

In an interview, Jimin is asked about what makes him so lovely and charming. He replies with "잘 모르겠어요 jal mo-reu-ge-sseo-yo," meaning he's not entirely sure. This expression combines "모르다 mo-reu-da" (do not know) and "잘 jal," which means "for sure" in this context. It's used when you have no idea or are uncertain about something. For example, if you're asked for directions on the street and you're not sure, you can say "잘 모르겠어요 jal mo-reu-ge-sseo-yo."

12

11

<Weverse Live> 2020.01.01

I just woke up.

방금 일어났어요.

bang-geum i-reo-na-sseo-yo

During a live video stream, RM says, "방금 일어났어요 bang-geum i-reo-na-sseo-yo" to explain that he just woke up. "방금 bang-geum" is used when referring to a moment that occurred just seconds ago, so the past tense phrase "일어났어요 i-reo-na-sseo-yo" is used. If someone asks you when you are coming, you can say "방금 출발했어(요) bang-geum chul-bal-hae-sseo(yo)" to tell them that you have just departed, conjugating "출발하다 chul-bal-ha-da" (to depart) in the past tense.

01

18

I'm bored. Please come quickly.

<Run BTS!> Ep.112

January 18th

I'm bored.

심심합니다.

shim-shim-ham-ni-da

Jin is playing the role of host. He is waiting for the others, while they prepare for their challenge. As it is taking longer than expected, Jin says, "**심심합니다** shim-shim-ham-ni-da." He's expressing that he is bored with the situation of waiting. You can casually say "**심심해** shim-shim-hae" to your close friends. Try saying "**심심해** shim-shim-hae" or "**심심합니다** shim-shim-ham-ni-da" when you feel bored and have nothing to do.

12

12

This item has various colors.

<Run BTS!> Ep.128

December 12th

have various colors
색이 다양하다
sae-gi da-yang-ha-da

BTS takes turns describing an eraser in the Liar Game. Jung Kook says "**색이 다양하다** sae-gi da-yang-ha-da," conjugating "색 saek" (color) and "**다양하다** da-yang-ha-da" (to be various). When you want to explain diversity, you can say "**N (noun)이/가 다양하다** i/ga da-yang-ha-da." As ice cream has many different flavors (맛 mat), you can say "**맛이 다양하다** ma-shi da-yang-ha-da" to describe it.

* Nouns ending with a batchim use "이 i," and nouns ending without one use "가 ga."

01

17

I'm a bit surprised.

[EPISODE] BTS (방탄소년단)
2021 'DALMAJUNG' Shoot

I'm surprised.
놀랐어요.
nol-la-sseo-yo

Hanbok, traditional Korean clothing, usually has brightly colored patterns of red, yellow, and blue. So when SUGA wears a silver *hanbok* for the first time, he is surprised at how good the color looks. He says, "놀랐어요 nol-la-sseo-yo" to express his astonishment. If you are greatly surprised by BTS' new release, try saying "놀랐어 nol-la-sseo" or "놀랐어요 nol-la-sseo-yo."

<Weverse Live> 2020.09.12

It's been two months.

두 달 됐어요.

du dal dwae-sseo-yo

RM, who has been exercising regularly for two months, says, "**두 달 됐어요** du dal dwae-sseo-yo" on a live video stream. When indicating the length of time that has passed, you can say "(time) **됐다** dwaet-da." After the number, you should add "**분** bun" for minutes, "**시간** shi-gan" for hours, "**일** il" for days, "**달** dal" or "**개월** gae-wol" for months, and "**년** nyeon" for years. Let's determine how long it has been since you started studying Korean!

01

16

January 16th

I feel very sleepy.

너무 졸려요.

neo-mu jol-lyeo-yo

During a live video stream in the early hours after midnight, Jin, who has been awake for too long, expresses his desire to sleep, saying "너무 졸려요 neo-mu jol-lyeo-yo." "졸려요 jol-lyeo-yo" is used when you are so exhausted that your eyes are starting to shut. Jin also uses the expression "너무 neo-mu" to emphasize just how sleepy he is. Are you sleepy from staying up late? In that case, you can say "너무 졸려 neo-mu jol-lyeo" or "너무 졸려요 neo-mu jol-lyeo-yo."

12

14

[BTS VLOG] V I DRIVE VLOG

December 14th

H: 오늘 하루 길어요.
o-neul ha-ru gi-reo-yo

It's been an eventful day.

오늘 하루 길다.

o-neul ha-ru gil-da

V has had a productive day. Looking back on everything he did, he says, "오늘 하루 길다 o-neul ha-ru gil-da." If many things have happened, you may feel that a day (하루 ha-ru) has been especially long (길다 gil-da). And yet there is still plenty of time until the day ends. To express this in a polite form, "오늘 하루 길어요 o-neul ha-ru gi-reo-yo" can be used. After studying Korean, watching a movie, going shopping, and meeting friends, is the sun still up? Wow, 오늘 하루 길어요 o-neul ha-ru gi-reo-yo!

01

Oh my gosh!

깜짝이야!

kkam-jja-gi-ya

Oh my gosh, 22.

<Run BTS!> Ep.146

To avoid being caught by the "it," the director of <Run BTS!>, SUGA and Jung Kook hide in the set one after the other and are startled by a mannequin that looks like a person. At that moment, they shout, "깜짝이야 kkam-jja-gi-ya," which is an expression used when you are alarmed by something. After Jung Kook says it, SUGA uses the exact same words. They are written in the subtitles as "깜짝이야22 kkam-jja-gi-ya." The numbers imply that the thought has been seconded. If another member had said "깜짝이야 kkam-jja-gi-ya," there would have been a "33" following the expression in the subtitles.

12

December 15th

It's true.

진짜예요.

jin-jja-ye-yo

On a live video stream, Jimin says he misses ARMY and then adds "**진짜예요** jin-jja-ye-yo," which means "I mean it" or "It's true." This expression is used when you sincerely mean something and want someone to believe that what you are saying is true. You can casually say "**진짜야** jin-jja-ya" to your close friends. As you are doing a good job learning Korean with BTS, your Korean will improve soon. **진짜예요** jin-jja-ye-yo!

01

14

Hug suddenly

<Run BTS!> Special Episode
- Telepathy Part 1

H: 다행이에요.
da-haeng-i-e-yo

What a relief.

da-haeng-i-da

While BTS is on a mission to visit locations that bring back memories for them, V finds out that Jin and Jimin chose the same location as he did. He says "**다행이다** da-haeng-i-da" to show that he feels relieved. This expression is used when things turn out as planned and you feel relieved. Did you manage to get tickets for the BTS concert? Then try saying "**다행이다** da-haeng-i-da" or "**다행이에요** da-haeng-i-e-yo."

12

16

j-hope: Hyung, be careful, be careful.

<Run BTS!> Special Episode
- Fly BTS Fly Part 1

December 16th

H: 조심하세요.
jo-shim-ha-se-yo

Be careful.

조심해.

jo-shim-hae

After a flying yoga session, BTS is stepping down from the hammock! When Jin seems to lose his balance, j-hope says "**조심해** jo-shim-hae," which means "Be careful." To express this in a polite form, "**조심하세요** jo-shim-ha-se-yo" can be used. If a car is approaching when someone is crossing the street, you can say "**조심해** jo-shim-hae!" or "**조심하세요** jo-shim-ha-se-yo!"

13

It has been a while since he felt relaxed.

<Run BTS!> Ep.127

H: 여유로워요.
yeo-yu-ro-wo-yo

It's relaxing.

여유롭다.

yeo-yu-rop-da

j-hope finally completes all 14 missions and can now enjoy some free time. He says, "여유롭다 yeo-yu-rop-da," which means "It's relaxing." How long it has been since he got to relax! When you have some time off from work or school and are enjoying a relaxing moment, try using this expression: 여유롭다 yeo-yu-rop-da! 여유로워요 yeo-yu-ro-wo-yo!

12

17

이게 뭐야

What is this?

<Run BTS!> Special Episode
- 'RUN BTS TV' On-air Part 2

December 17th

H: 이게 뭐예요?
i-ge mwo-ye-yo

What is this?

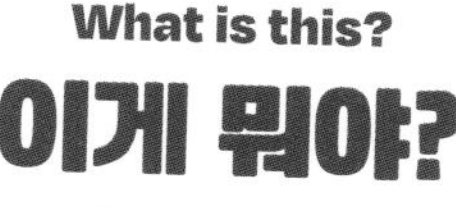

i-ge mwo-ya

Jimin opens the wrapper of a piece of gummy candy shaped like a fried egg on a pan and says, "**이게 뭐야** i-ge mwo-ya?" This can be literally used to ask what something is, but it can also be used to express embarrassment or confusion. When someone holds an item that you don't understand what it's for or when your pet messes up your room, you can try using this expression: **이게 뭐야** i-ge mwo-ya? **이게 뭐예요** i-ge mwo-ye-yo?

01

12

<Run BTS!> Ep.105

The '95s are peaceful as ever.

H: 평화로워요.
pyeong-hwa-ro-wo-yo

It's peaceful.

평화롭다.

pyeong-hwa-rop-da

As V decorates his clothes by splattering paint on them, Jimin approaches and spreads out his clothes as well. V decorates Jimin's clothes with a smile, and the two seem to be in a peaceful mood. At that moment, the phrase "**오늘도 구오즈는 평화롭다** o-neul-do gu-o-jeu-neun pyeong-hwa-rop-da" appears in the subtitles. In this expression, "**구오즈** gu-o-jeu" ('95s) refers to Jimin and V, who were born in 1995, and "**평화롭다** pyeong-hwa-rop-da," means "to be peaceful without any conflict." Jimin and V are a perfect example of the expression "**평화롭다** pyeong-hwa-rop-da," aren't they?

12

It's good for health.

건강에 좋습니다.

geon-gang-e jo-sseum-ni-da

BTS is having a discussion about whether to put eggs or peas on their *jjajangmyeon* (noodles in black bean sauce) in an episode of <Run BTS!>. j-hope, who loves peas, says, "**건강에 좋습니다** geon-gang-e jo-sseum-ni-da," meaning that peas are good for health (**건강** geon-gang). What else is beneficial to our body? Make your own answer like this: 운동 (exercise)은 건강에 좋습니다 un-dong-eun geon-gang-e jo-sseum-ni-da.

01

11

Certainty

<Run BTS!> Ep.147

January 11th

H: 확실해요.
hwak-shil-hae-yo

I'm sure.

확실해.

hwak-shil-hae

BTS has to find out which member is playing the role of a thief! j-hope says, "확실해 hwak-shil-hae," as he is confident that Jin is the thief. This expression is used when something is assumed to be true based on a guess. In honorifics, "확실해요 hwak-shil-hae-yo" can be used to make it sound formal and polite. Have your Korean skills improved today? 확실해요 hwak-shil-hae-yo!

19

Hmm... What else could there be...

<Run BTS!> Ep.124

H: 뭐가 있을까요?
mwo-ga i-sseul-kka-yo

What else could there be?
뭐가 있을까?
mwo-ga i-sseul-kka

BTS is brainstorming new ideas for <Run BTS!>. Jin mumbles to himself, saying, "**뭐가 있을까** mwo-ga i-sseul-kka?" This expression is used when asking others or talking to yourself about what other options there could be. When you ponder how to show BTS your heart, try saying this expression to yourself: **뭐가 있을까** mwo-ga i-sseul-kka?

<Run BTS!> Ep.154

We're curious!

We're curious.

궁금해요.

gung-geum-hae-yo

Around 200,000 ARMY voted in the Top Song Survey of <Run BTS!>. When the results are about to be revealed, BTS says "**궁금해요** gung-geum-hae-yo." This expression is used when you wonder about something. You can hear this expression in BTS song <Boy With Luv>. The line "**모든 게 궁금해** mo-deun ge gung-geum-hae" translates to "I'm curious about everything." What is your favorite BTS song? **궁금해요** gung-geum-hae-yo!

12

20

<Run BTS!> Ep.127

Please wait.

기다리세요.

gi-da-ri-se-yo

Jung Kook is about to go home after finishing his mission, and Jimin asks him to wait by saying "**기다리세요** gi-da-ri-se-yo." This is an imperative sentence used when directly telling someone to wait. You can simply say "**기다려** gi-da-ryeo" to your close friends. However, keep in mind that it is impolite to use imperative sentences with people older than you in Korea!

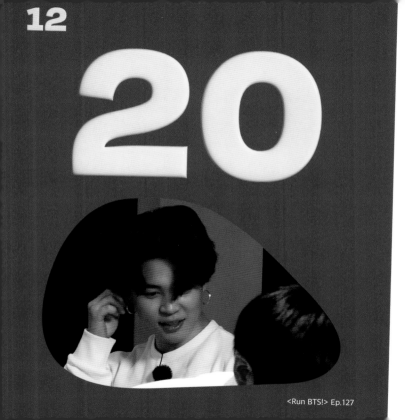

01

RM
명절 음식은... 역시 저는 잡채 제일 좋아해요

Holiday dishes... *Japchae* is definitely my favorite.

[EPISODE] BTS (방탄소년단) 2021 'DALMAJUNG' Shoot

Japchae is my favorite.

잡채를 제일 좋아해요.

jap-chae-reul je-il jo-a-hae-yo

In Korea, there are a variety of traditional dishes eaten during holidays. RM expresses his love for *japchae* (stir-fried glass noodles and vegetables) by saying "**잡채를 제일 좋아해요** jap-chae-reul je-il jo-a-hae-yo." When something is your favorite, you can say "**N (noun)을/를 제일 좋아해요** eul/reul je-il jo-a-hae-yo." In this phrase, "**제일** je-il" means "the most." When someone asks you who your favorite music artist is, try this expression: **방탄소년단을 제일 좋아해** bang-tan-so-nyeon-da-neul je-il jo-a-hae, **방탄소년단을 제일 좋아해요** bang-tan-so-nyeon-da-neul je-il jo-a-hae-yo.

* Nouns ending with a batchim use "을 eul," and nouns ending without one use "를 reul."

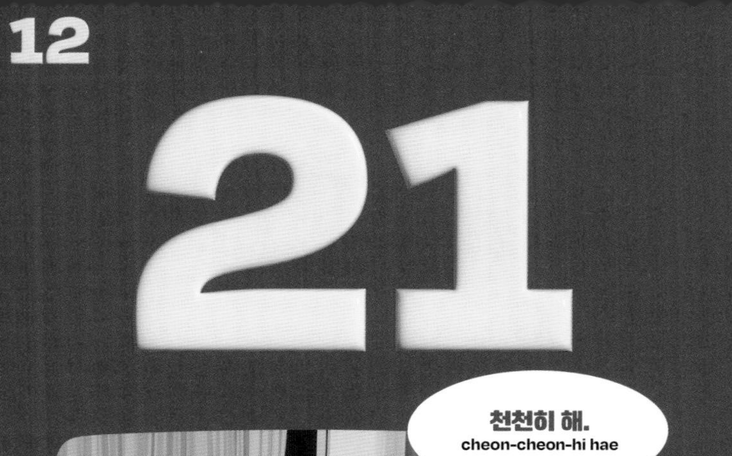

Coach entering

<Run BTS!> Ep.117

December 21st

H: 천천히 하세요.
cheon-cheon-hi ha-se-yo

Take your time.

천천히 해.

cheon-cheon-hi hae

While playing a water-bottle tossing game, RM tells the others "**천천히 해** cheon-cheon-hi hae," which means "Take your time." "**천천히** cheon-cheon-hi" means "slowly." In honorifics, you can say "**천천히 하세요** cheon-cheon-hi ha-se-yo" to make it more formal and polite. If someone is rushing even though it's not an urgent situation, you can say "**천천히 해** cheon-cheon-hi hae" or "**천천히 하세요** cheon-cheon-hi ha-se-yo."

01

08

기대가 됩니다.
gi-dae-ga doem-ni-da
사회자 진은
금지어로 어떤 말을 적었을지?

What would the host Jin have written as a restricted word?

<Run BTS!> Ep.131

I'm excited.

기대가 됩니다.

gi-dae-ga doem-ni-da

BTS is playing a game in which they are penalized each time they say or do certain things! Jin has to choose either a restricted word or a restricted action, and j-hope says "기대가 됩니다 gi-dae-ga doem-ni-da" because he is so excited to find what Jin decided. "기대(가) 되다 gi-dae(ga) doe-da" means "to look forward to something and feel elated." If you are excited about a trip to Korea next week, try saying "기대(가) 돼 gi-dae(ga) dwae" or "기대(가) 됩니다 gi-dae(ga) doem-ni-da" to show your excitement.

12

22

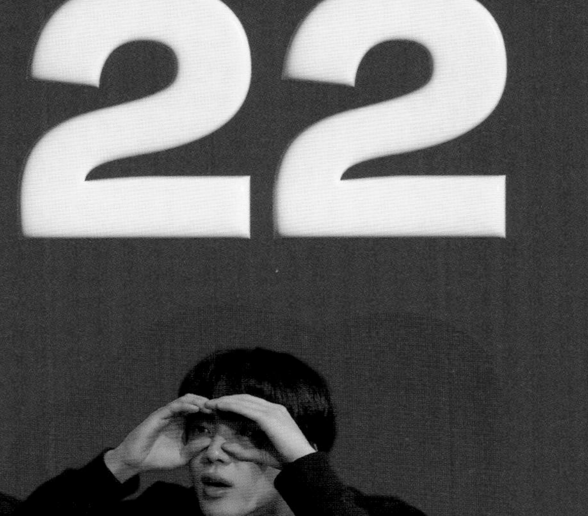

<Run BTS!> Ep.101

one more time

한 번 더

han beon deo

Everyone except V must guess what's inside a metal box that V has opened and closed quickly. They ask him to show it one more time, saying, "**한 번 더** han beon deo." This means "one more time" and can be used to ask for something to be repeated. If you are watching a BTS music video with a friend and want to replay it, you can say "**한 번 더** han beon deo!"

01

Jin
저는 이미 완성이 돼있어 가지고 매우 만족스럽습니다

I'm already complete, so I'm very satisfied.

[EPISODE] BTS (방탄소년단) 2021 'DALMAJUNG' Shoot

January 7th

I'm very satisfied.

매우 만족스럽습니다.

mae-u man-jok-seu-reop-seum-ni-da

There is a saying that fashion is completed with the face! Jin, who is handsome and looks great in anything he wears, expresses his satisfaction with his situation by saying "매우 만족스럽습니다 mae-u man-jok-seu-reop-seum-ni-da." "만족스럽습니다 man-jok-seu-reop-seum-ni-da" means that you are happy with the situation, and "매우 mae-u" is added for emphasis. If you secure a seat with a good view at a BTS concert, try saying "매우 만족스러워 mae-u man-jok-seu-reo-wo!" or "매우 만족스럽습니다 mae-u man-jok-seu-reop-seum-ni-da!"

12

23

December 23rd

It's at home.

집에 있어요.

ji-be i-sseo-yo

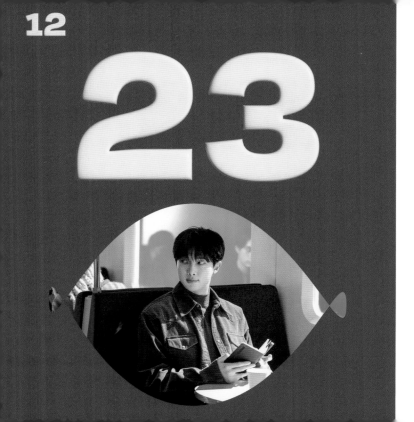

On a live video stream, RM is asked where the doll that resembles him is. He replies, "집에 있어요 ji-be i-sseo-yo," which means "It's at home." The phrase "N (noun)에 있어요 e i-sseo-yo" implies that something is in a certain place. You can use this expression by replacing "N (noun)" with a variety of places. For example, when you're at home and someone asks where you are, you can say "집에 있어 ji-be i-sseo" or "집에 있어요 ji-be i-sseo-yo."

01

[2021 FESTA] BTS (방탄소년단)
'방림이네 사진관 (Mission! BTS 4 Cuts)'

January 6th

I really like it.

굉장히 마음에 듭니다.

goeng-jang-hi ma-eu-me deum-ni-da

j-hope is very happy with the frame that holds a picture taken with V and says "**굉장히 마음에 듭니다** goeng-jang-hi ma-eu-me deum-ni-da." "**마음에 들다** ma-eu-me deul-da" means that something pleases one's heart, indicating a high level of satisfaction. Moreover, "**굉장히** goeng-jang-hi" (to a great extent) implies that he must have been very happy with the photo frame. If you love some beautiful new clothes, try saying "**굉장히 마음에 들어** goeng-jang-hi ma-eu-me deu-reo" or "**굉장히 마음에 듭니다** goeng-jang-hi ma-eu-me deum-ni-da" to express your satisfaction.

12

24

It's late.

늦은 시간이에요.

neu-jeun shi-ga-ni-e-yo

j-hope goes live to celebrate his birthday. When he realizes it's 12:30 a.m. in Korea, he says, "늦은 시간이에요 neu-jeun shi-ga-ni-e-yo," which means "It's late." While talking with your friend about BTS and enjoying the conversation, if you realize that it's already past midnight, try saying "늦은 시간이야 neu-jeun shi-ga-ni-ya" or "늦은 시간이에요 neu-jeun shi-ga-ni-e-yo" and suggest continuing the conversation the next day.

It was fun.

재밌었어요.

jae-mi-sseo-sseo-yo

It was so much fun. BTS (방탄소년단) RM's BE-hind 'Full' Story

While talking about writing the lyrics for the song <Life Goes On>, RM and Jimin explain that they had a challenging but fun time by saying "**재밌었어요** jae-mi-sseo-sseo-yo." This expression is the past tense of "**재밌어요** jae-mi-sseo-yo," which means "It is fun." If you are doing something fun at this moment, try saying "**재밌어(요)** jae-mi-sseo(yo)." If you have done something fun in the past, you can say "**재밌었어(요)** jae-mi-sseo-sseo(yo)" to express how fun it was!

Oh, I know this game.

<Run BTS!> Special Episode
- 'RUN BTS TV' On-air Part 1

H: 저 이 게임 알아요.
jeo i ge-im a-ra-yo

I know this game.

나 이 게임 알아.

na i ge-im a-ra

Jin is feeding a bear cake in a game, and Jung Kook says, "나 이 게임 알아 na i ge-im a-ra," meaning that he knows the game. When you want to tell someone that you know something, you can replace "이 게임 i ge-im" with whatever you want. For example, to say that you know Korean (한국어 han-gu-geo), you can say this expression: 나 한국어 알아 na han-gu-geo a-ra! 저 한국어 알아요 jeo han-gu-geo a-ra-yo!

01

04

Feels good? /
Feels good!

<Run BTS!> Ep.125

January 4th

H: 좋아요? 좋아요!
jo-a-yo? jo-a-yo!

Good? Good!
좋아? 좋아!
jo-a? jo-a!

While j-hope is cooking, he mashes ham inside a plastic bag and repeats the expression "느낌 좋아 neu-kkim jo-a!" which means "It feels good!" RM asks him if it feels good by saying "느낌 좋아 neu-kkim jo-a?" When you want to ask someone if something is good, try asking "N (noun) 좋아 jo-a?" If it is good, you can reply "N (noun) 좋아 jo-a!" 방탄소년단 좋아 bang-tan-so-nyeon-dan jo-a? 방탄소년단 좋아 bang-tan-so-nyeon-dan jo-a!

12

Wait a moment.

잠깐만요.

jam-kkan-man-yo

In an episode of <Run BTS!>, while listening to j-hope, RM suddenly shouts, "**잠깐만요** jam-kkan-man-yo" and then gives his opinion. This expression is used when you ask someone to wait a moment. You can casually say "**잠깐만** jam-kkan-man" to your close friends. If you need time to think or do other things first, try using this expression: **잠깐만** jam-kkan-man, **잠깐만요** jam-kkan-man-yo.

01

03

I trust you.

[2020 FESTA] BTS (방탄소년단) '방탄생파'

H: 전 당신을 믿어요.
jeon dang-shi-neul mi-deo-yo

I trust you.

난 널 믿어.

nan neol mi-deo

Jin and RM are making *kalguksu* (noodle soup) for the first time in their lives. Jin expresses his trust in RM by saying "난 널 믿어 nan neol mi-deo," which means "I trust you." In this expression, "널neol" (you) is short for "너를 neo-reul." To make the expression sound formal and polite in honorifics, you can say "전 당신을 믿어요 jeon dang-shi-neul mi-deo-yo." When you would like to show your trust in someone, you can say "난 널 믿어 nan neol mi-deo" or "전 당신을 믿어요 jeon dang-shi-neul mi-deo-yo."

See you at 12 o'clock.

<Run BTS!> Ep.136

H: 12시에 봐요.
yeol-ttu-shi-e bwa-yo

See you at 12 o'clock.

12시에 봐.

yeol-ttu-shi-e bwa

You can say, "(time)에 봐 e bwa" when asking to meet someone at a particular time. Jin says, "12시에 봐 yeol-ttu-shi-e bwa," which means "See you at 12 o'clock." Of course, you can specify the minute by saying "6시 13분에 봐 yeo-seot-shi ship-sam-bu-ne bwa," which means "See you at 6:13." Try saying "(time)에 봐 e bwa" or "(time)에 봐요 e bwa-yo" when you make an appointment!

<Run BTS!> Ep.151

work hard

힘차게 달리다

him-cha-ge dal-li-da

After enjoying a hotel staycation in an episode of <Run BTS!>, V uses the expression "힘차게 달리다 him-cha-ge dal-li-da" to mean that he'll work hard again in the future. This expression literally means "to run vigorously" and is a metaphor for working hard or doing something with enthusiasm. So, the expression "힘차게 달리다 him-cha-ge dal-li-da" can describe anyone who studies Korean hard every day, like you!

28

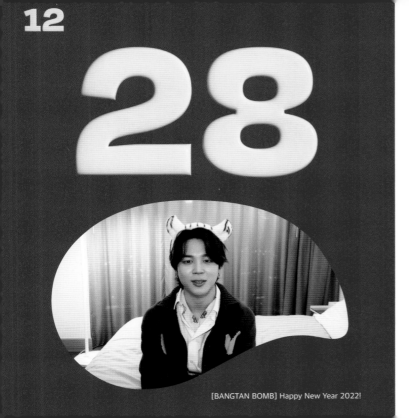

[BANGTAN BOMB] Happy New Year 2022!

Always stay healthy.

항상 건강하세요.

hang-sang geon-gang-ha-se-yo

Jimin finishes his New Year's message with "항상 건강하세요 hang-sang geon-gang-ha-se-yo." This expression is used when you hope someone always stays healthy. It can also be used at the end of letters, emails, or when you're saying goodbye to someone. Everyone, 항상 건강하세요 hang-sang geon-gang-ha-se-yo!

01

01

Happy New Year!

새해 복 많이 받으세요!

sae-hae bok ma-ni ba-deu-se-yo

Have you ever heard BTS say "새해 복 많이 받으세요 sae-hae bok ma-ni ba-deu-se-yo" on New Year's Day? This expression is used when the new year (새해 sae-hae) is here, to wish for good fortune (복 bok) for others. In Korea, people say this expression to send their regards on the first day of the solar and lunar calendar. Everyone, 새해 복 많이 받으세요 sae-hae bok ma-ni ba-deu-se-yo!

12

29

Have a good night.

편안한 밤 되세요.

pyeo-nan-han bam doe-se-yo

When content creator V's live video stream is about to finish in an episode of <Run BTS!>, SUGA, one of the viewers, says, "**편안한 밤 되세요** pyeo-nan-han bam doe-se-yo" before saying goodbye. Commonly, this is used before going to bed and means that you wish for someone to have a comfortable night. There are other expressions such as "**잘 자요** jal ja-yo" or "**안녕히 주무세요** an-nyeong-hi ju-mu-se-yo" that you can use. Before going to bed tonight, try using this expression: **편안한 밤 되세요** pyeo-nan-han bam doe-se-yo.

January

30

Happy Birthday, V!

생일 축하해요, 뷔!

saeng-il chu-ka-hae-yo, V

How to read the date in Korean

When reading the date in Korean, read the month first and then the day, as in "6월 13일 yu-wol ship-sa-mil" (June 13th). If you want to include the year, read it in the order of year, month, and day, as in "2013년 6월 13일 i-cheon-ship-sam-nyeon yu-wol ship-sa-mil" (June 13th, 2013).

Month 월

For the month, add "월wol," which means "month," after the number corresponding to each month.

January	1월	i-rwol
February	2월	i-wol
March	3월	sa-mwol
April	4월	sa-wol
May	5월	o-wol
June	6월	yu-wol
July	7월	chi-rwol
August	8월	pa-rwol
September	9월	gu-wol
October	10월	shi-wol
November	11월	shi-bi-rwol
December	12월	shi-bi-wol

Day 일

For the day, add "일il," which means "day," after the number corresponding to each day.

1st	1일	i-ril
2nd	2일	i-il
3rd	3일	sa-mil
4th	4일	sa-il
5th	5일	o-il
6th	6일	yu-gil
7th	7일	chi-ril
8th	8일	pa-ril
9th	9일	gu-il
10th	10일	shi-bil
11th	11일	shi-bi-ril
12th	12일	shi-bi-il
13th	13일	ship-sa-mil
14th	14일	ship-sa-il
15th	15일	shi-bo-il
16th	16일	shim-nyu-gil
17th	17일	ship-chi-ril
18th	18일	ship-pa-ril
19th	19일	ship-gu-il
20th	20일	i-shi-bil
21st	21일	i-shi-bi-ril
22nd	22일	i-shi-bi-il
23rd	23일	i-ship-sa-mil
24th	24일	i-ship-sa-il
25th	25일	i-shi-bo-il
26th	26일	i-shim-nyu-gil
27th	27일	i-ship-chi-ril
28th	28일	i-ship-pa-ril
29th	29일	i-ship-gu-il
30th	30일	sam-shi-bil
31st	31일	sam-shi-bi-ril

12

31

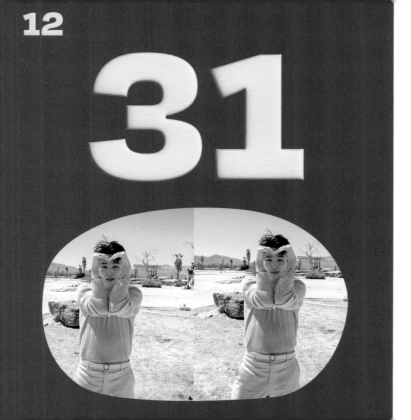

Let's make a lot of beautiful memories.

예쁜 추억 많이 만들어가봅시다.

ye-ppeun chu-eok ma-ni man-deu-reo-ga-bop-shi-da

On a live video stream, Jimin encourages ARMY to create many beautiful memories together, saying, "**예쁜 추억 많이 만들어가봅시다** ye-ppeun chu-eok ma-ni man-deu-reo-ga-bop-shi-da." "**예쁜 추억** ye-ppeun chu-eok" means "beautiful memory," "**많이** ma-ni" means "a lot," and "**만들어가다** man-deu-reo-ga-da" means "to make." Therefore, he suggests to ARMY that they should continue to make many beautiful memories together in the future. Why don't we also say this to BTS? **예쁜 추억 많이 만들어가봅시다** ye-ppeun chu-eok ma-ni man-deu-reo-ga-bop-shi-da!

Consonants 자음

The Korean language has 19 consonants. Some are basic consonants, and others are their counterparts. They are created by adding a stroke to or repeating the basic consonant. The former are called "aspirated consonants," while the latter are called "tense consonants" or "double consonants."

These are consonants with basic sounds.

Pronounce them with a burst of breath.

Pronounce them intensely from the throat, without a burst of breath.

Basic Consonants	IPA	Romanization	Aspirated Consonants	IPA	Romanization	Tense Consonants	IPA	Romanization
ㅂ	[p]/[b]	b	ㅍ	[pʰ]	p	ㅃ	[p′]	pp
ㄷ	[t]/[d]	d	ㅌ	[tʰ]	t	ㄸ	[t′]	tt
ㅅ	[s]/[ʃ]	s, sh	–	–	–	ㅆ	[s′]/[ʃ′]	ss, ssh
ㅈ	[ts]/[dz]	j	ㅊ	[tsʰ]	ch	ㅉ	[ts′]	jj
ㄱ	[k]/[g]	g	ㅋ	[kʰ]	k	ㄲ	[k′]	kk
ㅁ	[m]	m	ㄴ	[n]	n	ㄹ	[r]/[l]	r, l
ㅇ	–	–	ㅎ	[h]	h			

When ㅇ is the beginning consonant, it serves only as a placeholder.

Batchim 받침

In writing, batchim can have various consonants, but there are only 7 (ㄴ, ㄹ, ㅁ, ㅇ, ㅂ, ㄷ, and ㄱ) pronunciations.

Batchim	IPA	Romanization
ㄴ	[n]	n
ㄹ	[l]	l
ㅁ	[m]	m
ㅇ	[ŋ]	ng

Batchim	Token Sound	IPA	Romanization
ㅂ, ㅍ	ㅂ	[p]	p
ㄷ, ㅌ, ㅅ, ㅆ, ㅈ, ㅊ	ㄷ	[t]	t
ㄱ, ㅋ, ㄲ	ㄱ	[k]	k

When ㅂ, ㄷ, and ㄱ are batchim, they are pronounced without a release of air, unlike the p, t, and k sounds in the final position of English words such as "cu**p**," "do**t**," and "boo**k**."

Korean Expressions Calendar

Publication Date	2023.06.20
Publisher	이충희 Lee, Choong Hee
IP Business Dept. Lead	최영남 Choi, Young Nam
IP Content Team Lead	서수진 Seo, Su Jin
Edu Content Part Lead	이서진 Lee, Seo Jin 정희은 Jeong, Hee Eun
Project Manager	김혜리 Kim, Hye Ri
Marketing Manager	김근영 Kim, Kun Young
Global Biz Team	정석교 Jeong, Seok Kyo (Lead) 박성희 Park, Sung Hee
Prof. of Korean as a Foreign Language	허용 Heo, Yong (HUFS)
Participating Researchers	강소희 Kang, So Hee 김학연 Kim, Hak Youn
English Translations	LEXCODE Inc.
English Proofreading	Patrick Anthony Ferraro
Design & Editing	금종각 Golden Bell Temple Graphics

Cake Corporation (Publication Registration May 24, 2022 No.2022-000170)
Pangyo TechONE Tower 1, 9F, 131, Bundannaegok-ro, Bundang-gu, Seongnam-si, Gyeonggi-do, 13529, Korea
ISBN 979-11-90996-65-5

Cake is the new name for HYBE EDU.

How to pronounce Hangeul

See the tables below for the IPA (International Phonetic Alphabet) symbols corresponding to each of the Korean consonants and vowels. Use them as a reference on how the Korean pronunciation sounds.

Vowels 모음

The Korean language has 8 simple vowels and 13 complex vowels. The complex vowels are created by adding a stroke to a simple vowel or combining simple vowels.

Simple Vowels	IPA	Romanization
ㅏ	[a]	a
ㅓ	[ə]	eo
ㅔ	[e]	e
ㅐ	[ɛ]	ae
ㅗ	[o]	o
ㅜ	[u]	u
ㅡ	[ɨ]	eu
ㅣ	[i]	i

Combination of /y/ sound and simple vowels

Complex Vowels	IPA	Romanization
ㅑ	[ja]	ya
ㅕ	[jə]	yeo
ㅖ	[je]	ye
ㅒ	[jɛ]	yae
ㅛ	[jo]	yo
ㅠ	[ju]	yu

Combination of /w/ sound and simple vowels

Complex Vowels	IPA	Romanization
ㅘ	[wa]	wa
ㅝ	[wə]	wo
ㅞ	[we]	we
ㅙ	[wɛ]	wae
ㅚ	[we]	oe
ㅟ	[wi]	wi

ㅢ	[ɨy]	ui

* The vowels grouped together in yellow highlighting have very similar pronunciations in practice.

Preview

Date

The top two digits indicate the month, and the bottom two digits indicate the day.

06

14

아미 보라해~♥

Translation of the subtitles in the photo

ARMY, I purple you.

<Run BTS!> Ep.97

Video source of the photo

Date

June 14th

Honorifics (H) notation

Honorific expressions and pronunciation of the main expression, which is given in informal language (banmal)

H: 보라해요.
bo-ra-hae-yo

English translation, Main Korean expression, and Pronunciation

I purple you.

보라해.

bo-ra-hae

Explanation of the main expression

Do you know the origin of "보라해 bo-ra-hae," which BTS and ARMY use instead of "사랑해 sa-rang-hae?" (☛ July 11th) "보라 bo-ra" means "purple," and "보라해 bo-ra-hae" was created by V when he saw ARMY BOMBs covered with purple plastic bags at a fan meeting in 2016. Just like purple is the color at the end of a rainbow, V explained, the love and trust that BTS and ARMY have for each other will be there in th

You can also practice the main expressions of <365 BTS DAYS> on the Cake app!

App Download